A NEW ECOPHYSIOLOGICAL APPROACH TO FOREST-WATER RELATIONSHIPS IN ARID CLIMATES

ISBN 978-94-017-0587-5   ISBN 978-94-017-0585-1 (eBook)
DOI 10.1007/978-94-017-0585-1

A NEW ECOPHYSIOLOGICAL APPROACH TO FOREST-WATER RELATIONSHIPS IN ARID CLIMATES

I. GINDEL

Springer-Science+Business Media, B.V. 1973

CONTENTS

Introduction 1

The Environment 5

The Correlation between the Morphology, Anatomy and Physiological Properties of the Forest Plant and its Environment 9

Methodology in Forest Research 23

The Consumption of Soil Water by Trees 33

Absorption of Atmospheric Moisture by Woody Xerophytes 49

Irrigation of Woody Xerophytes with Atmospheric Water within the Desert 61

Amount of "Flowing" Dew and Mist per Unit Area 62

Concentration of Rainwater by Means of Plastic Sheets 63

Accepted Transpiration Concepts 65

Transpiration during the Season of Growth 69

Transpiration as a Function of the following Ecophysiological Factors:

A. Physical Evaporation, Temperature, Relative Humidity and Wind Velocity 75

B. Transpiration as a Function of Soil Moisture Depletion 84

C. Transpiration as a Function of Solar Energy, Evaporation, Soil Moisture and Leaf Area 85

D. Transpiration as a Function of Leaf Moisture 90

E. Transpiration as a Function of Photosynthesis 95

F. Transpiration as a Function of Cambium Activity 100

G. Transpiration and Growth 104

The Xeromorphic Properties of the Leaf and their Relationship to the Process of Transpiration 111

Transpiration Suppressants 117

Xerophytism 119

Discussion and Conclusions 121

Bibliography 133

INTRODUCTION

A shortage of water exists, not only in the arid regions of the world, but even in some moderately humid climates. This situation is a consequence of water requirements for agriculture and industry in amounts greater than the natural surplus. Even in Europe there is increased anxiety over the state of water reserves, and shortages are forecast for the near future if industry continues to expand. During the past 50 years in the United States, water use has increased about twice as fast as the rate of population growth, and shortages have already appeared in some places.

The need to conserve declining water resources which has become apparent over the last few decades has led several investigators to conclude that plants with a high rate of transpiration endanger water resources, and the growth of such plants must not be encouraged. Some think that trees withdraw more water from the soil than other plant species and evaporate it excessively through the stomata of leaves.

THORNTHWAITE and HARE (1955) explained transpiration on the same thermodynamic basis as evaporation, and calculated its rate, using DALTON's law or modifications thereof. In spite of the many past and present investigations into the problems of transpiration, the biological aspects of this essential process is still poorly understood.

It will be of interest here to quote a sentence written by MILLER (1938), which is still relevant today: "Although transpiration has been studied, perhaps more thoroughly than any other plant process, very little is yet known concerning the significance of this function in the life of the plant". According to KOZLOWSKI (1964), the role of transpiration in plant growth and development has been a matter of vigorous controversy for many decades.

Another opinion criticizes the "speculative" approach of this process and its controlling factors, particularly the lack of reliable information for woody plants (F.A.O., 1962).

The orthodox interpretation of transpiration has led to erroneous interpretations of the morphology of the leaf—the essential pathway of the water vapour leaving the tree. In the case of woody xerophytes in arid climates there has been a lack of information concerning the role of atmospheric moisture in their water balance.

The conclusions that trees use more water than other plants may or may not be true when considering trees whose roots are in touch with the table water. In humid climates and in moist soils, high relative humidity and frequent rains often blur the boundaries between evaporation and transpiration. But the situation is altogether different for the conditions prevailing in arid climates with regard to the water regime of the woody plant grown without interference of man.

If from 10 to 40% of rain is intercepted by the forest canopy and evaporates before it reaches the forest floor, and if trees transpire more than other cover types,

1

then less moisture should remain in the forest soils after the rainless growing season than in the soils of neighboring treeless areas where there is only seasonal vegetation growth. The data from the arid zone contradict this supposition.

The relationship between water consumption by trees and the water regime of the ecosystem is little understood in these areas. Indeed, McGINNIES et al. (1968) confirms this when he states: "One of the main climatological problems of arid lands is the lack of a satisfactory system for determining water need and water use and for expressing aridity in these terms".

Only the ecology of plants grown within their native habitat or domesticated, without assistance of man, can reflect the natural correlation between the environment and the plant. This was a guiding principle in these investigations carried out in the desert and subtropical zone of Israel.

Within the climatic extremes, the ecological regime differs conspicuously in accordance with the quantity of water in the soil. In these studies, locations were selected in which plant roots were not in touch with a ground-water table.

The studies of plant-water relations under arid conditions led to unorthodox results and indicated the need to change our approach concerning a number of basic biological characteristics from the morphological, anatomical and ecophysiological points of view.

Woody plants growing under arid conditions, including the desert region where the annual rainfall ranges from 25 mm near the Red Sea to 200 mm at Beersheba, serve as interesting media for studying plant-water relations. This is especially so for plants growing without a reserve of soil moisture and which exist only on rainfall and atmospheric water (dew, mist, water vapor). In our ecological and morpho-anatomical investigations, we analyzed the following problems:

1. The relation between anatomo-morphogenesis and the environment.
2. The correlation between physical environmental conditions and biological properties of the plant.
3. The consumption of moisture by plants and the role of atmospheric moisture in the water balance of the plant.
4. The definition of the transpiration process as a function of other physiological processes.
5. The stomatal system and its ecophysiological role.
6. Radiation and its biological function.

It will be shown that the term "evapototranspiration", insofar as it engenders the supposition that transpiration and evaporation are identical physical processes, is a misnomer. In fact, evaporation has much in common with transpiration as it has with photosynthesis, rate of growth, or other metabolic activities. It will be shown that life phenomena of a tree cannot be reduced to physical definitions, but that they differ in accordance with the biological properties of species and varieties.

It will be shown that stomatal behavior is a very complicated biological phenomenon, keyed to the needs of the plant rather than to an "atmospheric demand". It follows that a fundamental revision is necessary in regard to the role of stomata and their interactions with the environment.

2

My deep appreciation is extended to Prof. PH. STOUTJESDIJK for his efforts in reviewing the MS; to Prof. A. HALEVY for reading the MS and for his useful remarks. My hearty thanks to the professional colleagues of the J.N.F., particularly to S. WEITZ and M. KOLAR, with whom I have worked for many years in harmony and who have given me the necessary technical assistance in the different locations of the desert and subtropical region of Israel in carrying out my investigations.

I. GINDEL

THE ENVIRONMENT

A survey of 28 volumes published by UNESCO revealed that their definition of "arid regions" is inconsistent from the ecological point of view, and often includes areas which are relatively mesophytic. It is therefore essential to define the arid conditions in which I worked before presenting the results obtained.

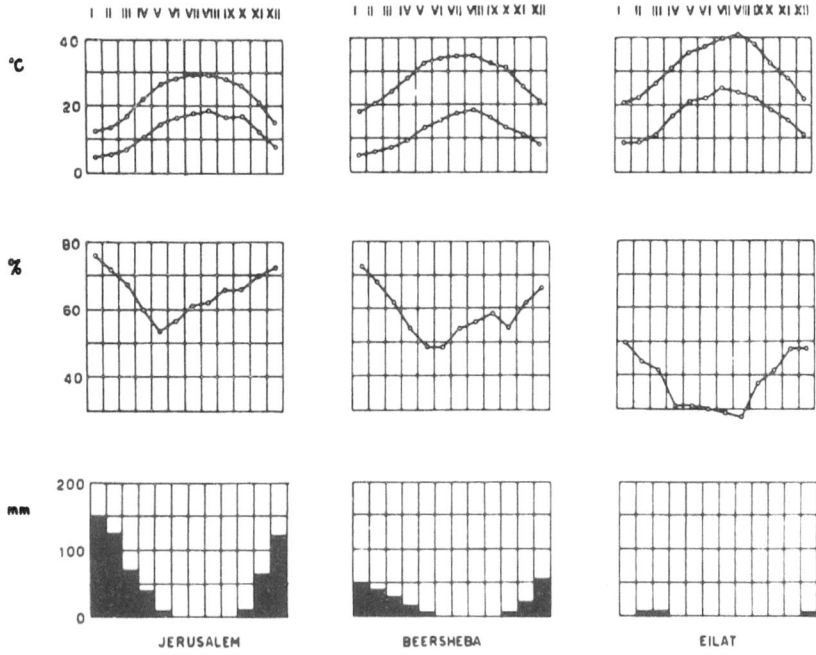

Fig. 1. Mean monthly temperature (maximum and minimum), relative humidity and rainfall at 3 localities belonging to 3 phytogeographic regions: Jerusalem–Mediterranean maqui, Beersheba–Irano–Turanian Zone and Eilath–Saharo–Sindian Zone.

Research was carried out in three climatic zones: subtropical, semi-desert, and desert. These three zones coincide with the three following isohyets: greater than 40 cm, from 20–40 cm, and from 2.5–20 cm, respectively (Fig. 1). The subtropical zone is often named Mediterranean, and is similar to those of California, Chile, Southern Australia, and South Africa, where the rainy season is the cool season.

No simple boundaries can be drawn between the zones. Microecological conditions significantly change the composition, morphology, and micro-structure of the vegetation. Differences are particularly striking between the southern and northern exposures.

The most extreme climatic conditions are found in the southern part of the desert: very high temperature: 42–47°C in the shade, accompanied by low relative air humidity (15–40%), and irregular yearly rainfall are the main features of this area. The mean annual rainfall is 7.5 cm falling during 15 days. Nevertheless, a natural woody vegetation is found along slopes and in local valleys where soil moisture is increased by the gravitational water. As a rule, soil moisture increases with depth from 1 to 4% in the upper layers to about 12% at a depth of 2 meters.

A correct description of the micro-climate can be given by a calculation in which all the important climatic factors are included (GINDEL 1957). The following formula is therefore proposed:

$$\text{Index of Aridity} = \frac{R \times N}{E}, \text{ where}$$

R = Yearly rainfall
N = Number of rainy days
E = Evaporation

$R \times N$ indicates the distribution of winter rainfall, which is of fundamental importance to growth: 400 mm of rainfall properly distributed is of far greater value than stormy rains concentrated during a shorter period (even if yearly rainfall is higher). Evaporation (E) includes the period of low relative humidity, the result of both high temperature and dry winds. By means of the above formula, therefore, the Index of Aridity can be calculated for the entire year, or for any particular period within the year.

The following phenomena modify the effects of precipitation on tree growth:

1. The variability in quantity of rainfall and its temporal distribution (these are most important in the semi-desert and desert).
2. The years of drought which occur often in all three zones, but particularly in the semi-desert and desert.
3. The rate of evaporation which reaches its peak during the season of the hot desert winds (Hamsin) when temperatures may rise by from 5.50°–10°C, and evaporation intensities may double or treble. The rainfall which runs from the slopes into the valleys and plains and remains on the surface of the saturated soil is largely depleted by evaporation. During clear days when soils are saturated their rate of evaporation equals that of open water. The continuous and intensive evaporation for many days between rains, and during the rainless hot season, is the most significant factor in arid regions in depletion of rainwater.
4. The difference between these regions in Israel and similar arid zones in other parts of the world is that during the rainy season temperatures are lowest, so that plant physiological activity is slowed or stopped. During the cold periods deciduous tree species shed their leaves, whereas the evergreens may be active

to some degree if absolute minimum temperatures do not drop below about 7°–8°C.

This situation is in contrast to other arid regions where the same quantity of rain falls during the warm season when the trees are in full growth. According to SLATYER (1965), in the areas where *Acacia aneurea* grows naturally, the average annual rainfall is 10″ of which 75 % falls during the summer months (October–March), and only 25 % during the winter months. In these conditions rainwater is exploited by the tree in a much more effective manner than in the Mediterranean Basin. Here the indigenous evergreens, for example Aleppo pine, Cypress sp., and others, start cambium activity in March or April, i.e., at the end of the rainy season.

5. Topography is very important from the ecological point of view. Only a small fraction of rain penetrates the soil when it falls on sloping ground where thin soils cover impermeable rock.

This is in contrast to the situation in level areas and depressions. Differences in the development of tree species planted in the following edaphic categories are striking: 1. Locations where the ground water is near the surface, and the roots are in contact with it. 2. Areas where, in addition to rainfall, soil moisture is increased by gravitational water from adjacent sloping areas. 3. Slopes, plains, and elevated areas, where the trees have only the capillary water available within the root zone.

The hilly country was well-wooded in ancient times, and due to the slow flow of gravitational water from the slopes of the well-forested mountains into the desert, the density of desert flora was greater. Following the destruction of the forests, erosion continued for centuries, and surface runoff carved out well-established beds (Wadis) leading towards the Mediterranean, Red and Dead Seas. Today, after a rainy period in the highlands, the flood-waters sweep down as veritable torrents, capable of carrying away fully loaded trucks (GINDEL, 1969).

The forest of the hilly subtropical region of Israel was destroyed beginning at the time of the occupation of the country by Rome in 63 B.C. Up to the period of the later Ottoman Empire (GINDEL, 1944, 1952, 1967); destruction was particularly severe by Bedouin tribes. The trees were not only cut for firewood, but also eradicated for manufacture of high quality charcoal from the roots.

The remnants of the natural woody vegetation persisted only by coppicing. After removal of the forests, litter and soil were washed away along the slopes by rains or blown by winds during the dry season. The hilly region is now entirely soil-less, or else has a shallow layer of 1 foot or less, or rarely 2 or 3 feet.

In the hilly and sloping Mediterranean region, the Cenomanean and Turanian geological formations are composed of hard limestone which is the parent material of terra rossa. The soft limestone middle and upper Senonian, and the soft rocks of Eocene, are the sources of gray soil often of the rendzina type.

On these soils the remnants of the Mediterranean forest consists of sclerophyllous trees and shrubs, mixed with evergreen and deciduous species. The main tree species are: Quercus calliprinos, Q. infectoria, Q. ithaburensis, Pinus halepensis, Cupressus horizontalis, Pistacia lentiscus, P. palaestina, P. atlantica, Ceratonia siliqua, Prunus ursina, and Acer syriacum. Characteristic shrub or small tree species are: Arbutus andrachne, Phillyrea media, Cercis siliquastrum, Myrtus communis, Rhamnus sp., Rhus coriaria, Spartium junceum, and Styrax officinalis.

The Irano–Turanian zone represents the semi-desert region. It consists mainly of herbs and semi-shrubs and of a small number of tree and shrub species, for example, Pistacia atlantica, Zisyphus spina-Christi, and Retama roetam. The plains of this zone are composed of loess, soil originated by winds from southern and eastern deserts. It is mainly composed of fine sand with a high percentage of lime.

The Saharo–Sindian, the desert zone, supports a small number of woody species. The dominants from the silvicultural point of view are Acacia tortilis, A. albida, A. spirocarpa, Calligonum commosum, Atripiex halimus, Anabasis articulata, Suaeda sp., Haloxylon persicum, Retama roetam, and Callotropis procera: All these species grow naturally in the Sahara Desert.

THE CORRELATION BETWEEN THE MORPHOLOGY, ANATOMY AND PHYSIOLOGICAL PROPERTIES OF THE FOREST PLANT AND ITS ENVIRONMENT

The mechanism of life is a multiplicity of biochemical reactions with enzymes and hormones involved in the conversion of starch to sugar (and vice versa) or to other hydrolytic products of high osmotic activity which form a source of kinetic energy. The phenomenon of life is a continuous metabolism which is strongly influenced by environmental factors. The environment is the initiator, sustainer, and regulator of plant life. An endemic or domesticated woody plant, grown naturally without the intervention of man, reflects its environment (i.e., knowing the environment makes it possible to define the pattern of growth). The following sections are an analysis of the observations of life phenomena in tree xerophytes in arid surroundings.

BILLING's (1952) diagrammatic representation of the interrelationship of environmental factors and plants demonstrated the complexity of various stages of the metabolic chain. Interference with one link in the chain affects the others. Unfortunately, little quantitative information is available regarding the internal correlation of life processes. It is evident that an examination of a single, steady-state physiological process which ignores simultaneous interacting processes does not reveal the general dynamic nature of plant life.

More than fifteen ecological factors act simultaneously on the life of the plant. The dynamic action of these factors is accompanied by an evolutionary process, or a process of adaptation of the plant to its environment. The evolution is evidenced in modification of plant morphogenesis, anatomy, and physiological functions.

Among the 15 or more environmental factors affecting plant life, light, temperature and water have the most obvious effects on physiological functions and adaptation phenomena. We shall term these three factors the "hydro-photo-thermal constellation". Changes in individual factors, or changes in the complex as a whole, affect the structure of plant cells and tissues. During certain parts of the growing season, cells are formed with large pores and thin walls; at other times the formation of mechanical tissues lacking vessels is emphasized, or canals may be formed which concentrate resin, or tannins, etc.

The plant exists in complete harmony with its environment. The structure of the various tissues and their relation to one another are a reflection of the environmental influence, and fluctuations in the morphology of the plant correspond to differences in the hydro-photo-thermal constellation. These facts are especially apparent in native or adapted plants growing without human interference.

9

Fig. 2. Yearly variations in ring-pattern of a dominant Aleppo-pine in a 48-year old dense plantation (Mt. Carmel).

Changes in tree structure caused by fluctuations in the environmental complex affect simultaneously many of its limbs.

The pattern of growth reflects the pattern of the climate, and the more heterogeneous the climate, the more heterogeneous is the pattern of growth (Fig. 2). This is also clearly shown by the structure of the xylem in tree trunks (Figs. 3–6). In a

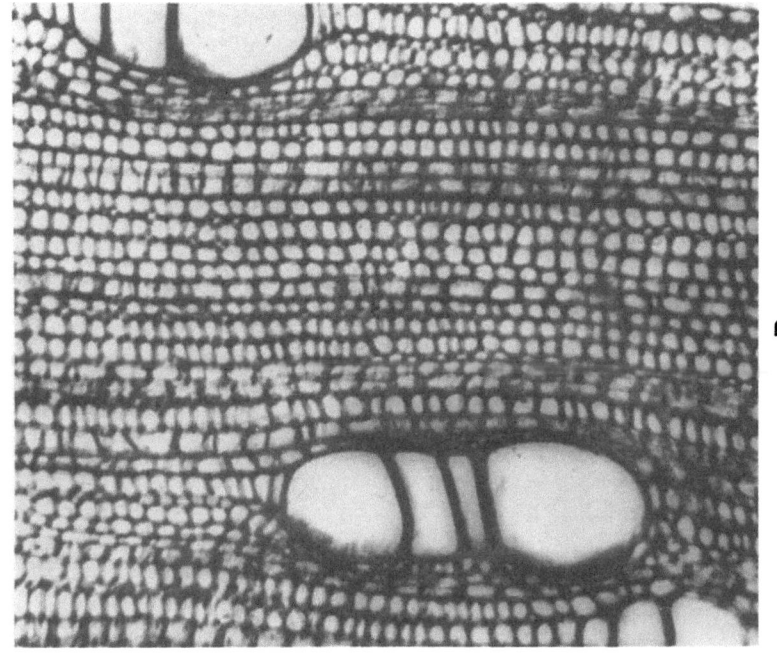

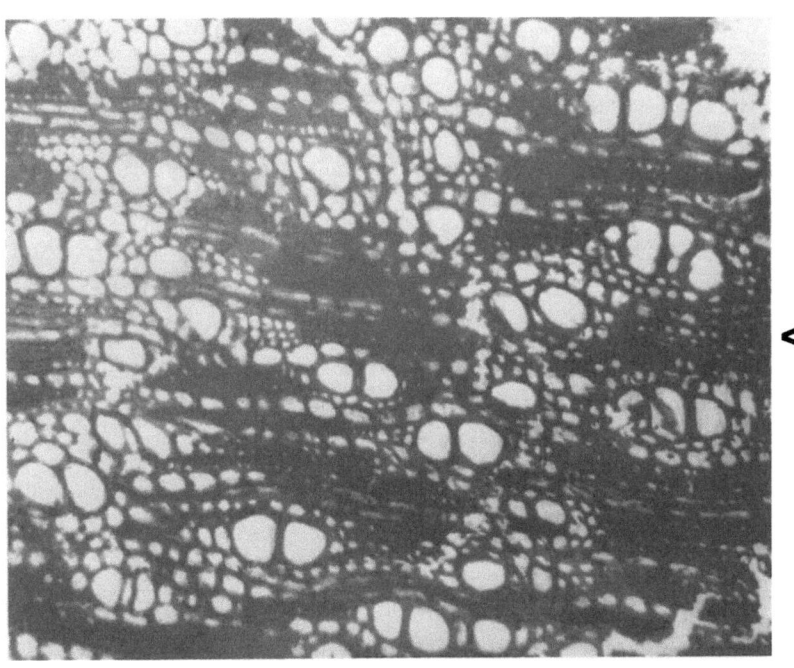

Fig. 3. Evergreen-diffuse-porous: A. *Olea europea,* and B. *Ricinus communis.*

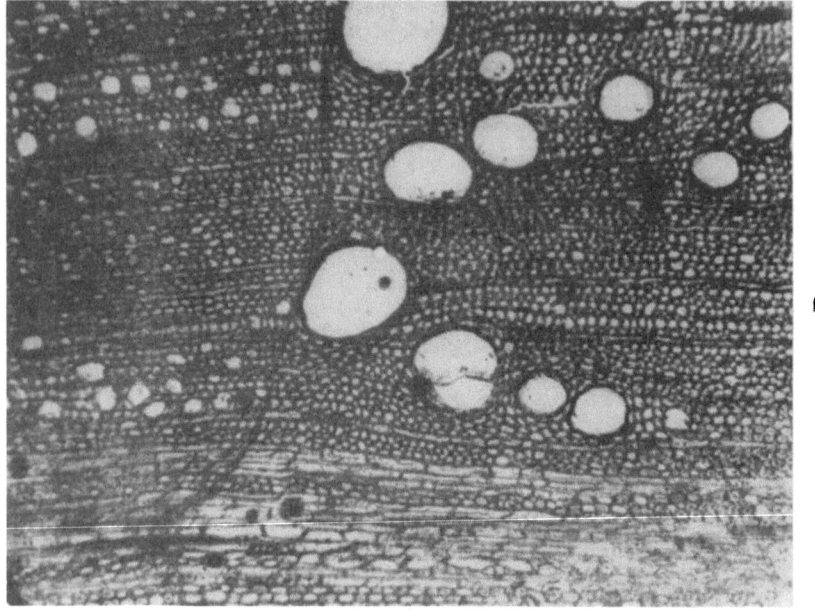

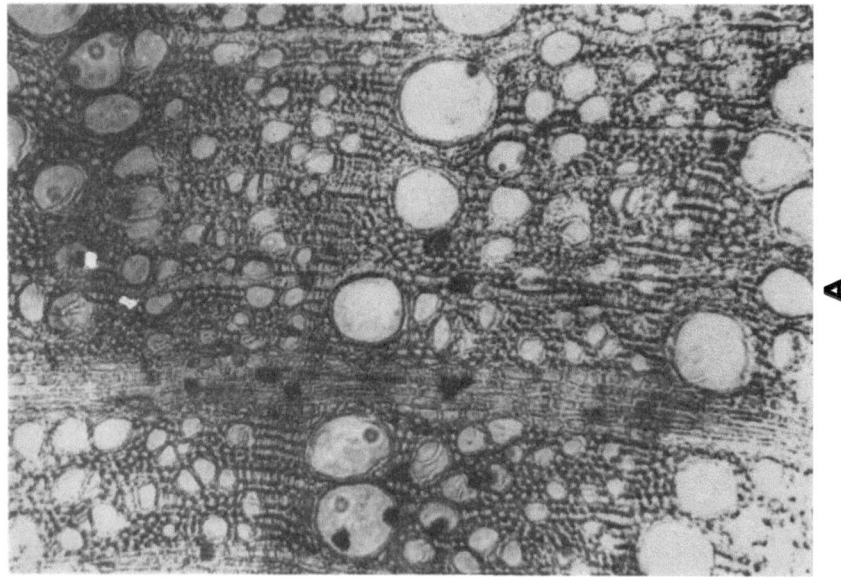

Fig. 4. Deciduous-ring-porous: A. *Amygdalus communis*, B. *Quercus ithaburensis.*

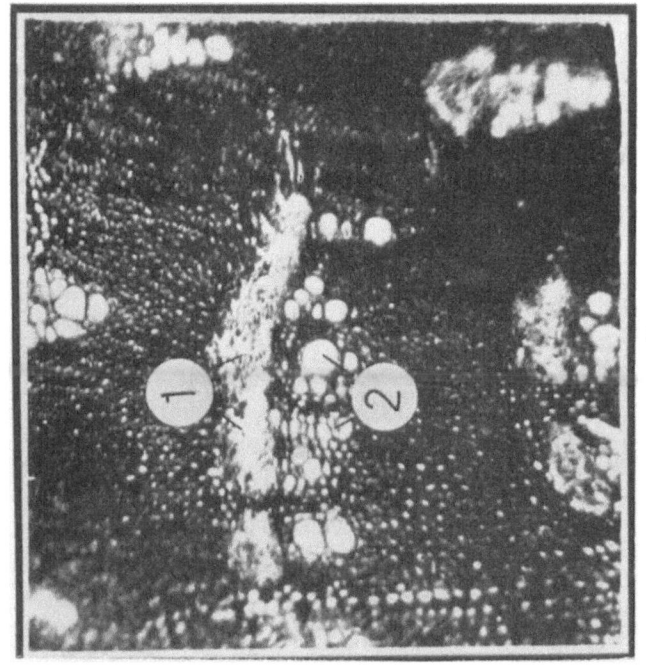

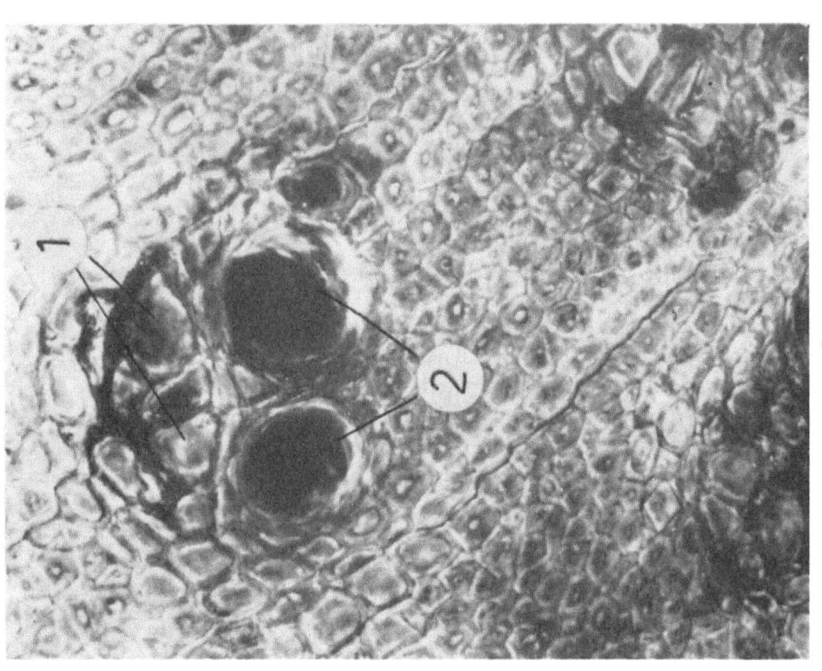

Fig. 5. A. -*Atriplex halimus.* L. B. -*Salsola tetrandra* 1. -phloem. 2. -xylem.

very homogeneous climate, with negligible changes in the constellation of eco-logical factors during the year, e.g., in the tropical rainforest, very homogeneous growth occurs which is easily recognized by its structure. The same species in a variable climate show heterogeneity of growth, as we found in exotic tree species transferred from a wet, tropical climate to the subtropical semi-arid zone.

In the desert region, where from 25 to 200 mm of rain falls annually, many plant species show unusual structures. The abnormality is expressed in *Atriplex, Eurotya,*

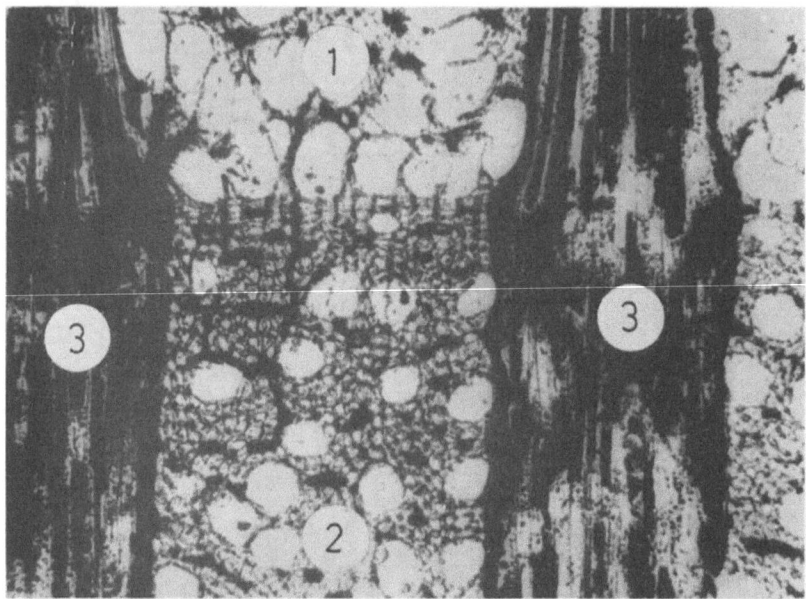

Fig. 6. Hygrophytic species: *Platanus orientalis.* L. 1. Spring wood. 2. Summer wood. 3. Medullary rays.

Graya, Salsola, and *Simmondsia* by the presence of phloem that has actually sunk within the xylem (Fig. 5). The ecological reasons for this structure, which is typical of desert regions, have not been elucidated.

The hygrophytic vegetation, growing beside permanent streams, responds to temperature and light, but is naturally not affected by a water deficit in the soil. Nevertheless, the hygrophytes show a characteristic decrease in the diameter of vessels in summer. Likewise, the walls of the primary tissue in which the vessels are embedded are thicker in the dry summer even though water is available in the soil all the year round (Fig. 6). In addition, while the xerophytes develop vessels of a relatively small area, in the hygrophytes the vessels occupy maximal area, as it can be seen in typical hygrophytes, like *Nerium oleander, Platanus orientalis, Populus euphratica, Salix* sp. and *Vitex agnus-castus.*

Arid climatic conditions affect the xylem of the indigenous vegatation in the following ways (GINDEL, 1952): a. small or medium-sized pores (diameters exceed 200–300 microns in only a few tree species); b. the presence of crystals, especially towards the periphery of the annual ring; c. more than 85 % of evergreen species are diffuse-porous (Fig. 4), and roughly 2·5 % resemble more closely the ring-porous type. The majority of the deciduous species are ring-porous. Intermediate wood structure is to be found in about 35 %. This characteristic makes it possible to divide the native vegetation into two classes. In the one, the form of dispersal of the tracheids is preserved but their diameter decreases as summer approaches and reaches a minimum at the periphery of the annual ring. In the other, the dispersal of the tracheids changes with the season.

The morphology and anatomy of a tree are correlated with its physiology, and all three reflect environmental influences. The different anatomical structure of the evergreens (e.g., diffuse-porous wood) is connected with a different physiologic mode of life, and characteristic habitat. The rate of transport of fluids in evergreens varies between 1 and 6 meters per hour, while in ring-porous deciduous trees it varies from 25 to 60 meters per hour (BAUMGARTEN, 1934).

Variations in ecological conditions effect not only morphological, but also chemical changes. *Eucalyptus camaldulensis*, for example, growing in the more rainy subtropical regions of the country (Hedera), has a different chemical constitution from the same *Eucalyptus* sp. in the desert. In the dry area it has higher content of non-resistant pentosans and hexosans, higher hot water and alkali solubility, lower alpha cellulose and higher acetyl groups and uronic anhydride content (LEWIN, 1956). Alkali solubility refers to soluble salts, low molecular weight sugars and polysaccharides-pentosans, gamma and beta cellulose.

The tannin content varies in accordance with direct illumination on the bark. With the apple and pear trees, which in some years bear fruit twice, as at Rehovot, it has been found that taste and size of the second crop, harvested in November, differs so much from those of the first crop harvested in July, that they might be fruit of two different varieties.

An illustration of the effect of climatic variations on its structure has been obtained from an examination of the changes effected in some of the 600 species of exotic trees brought into Israel from wet tropical climates, and cold climates. Figure 7 shows, for example, two sections through the trunk of two 4-year old coffee trees of the same variety (Bourbon). One grew in Tanganyika, on the slopes of Mount Kilimanjaro, where there is precipitation throughout the year and a minimum temperature of about 17°C. The other grew in sandy soil at Rehovot, assisted by irrigation and manuring. There is an obvious difference in the woody structure of the two trees. Section 2, from Israel, shows heterogeneity in the diameter of the vessels, and the small vessels at the beginning of the season of growth enlarge later to a maximum. At the end of the season, a strip of mechanical tissue is formed, clearly differentiating between the annual rings. The strip is completely absent in the section from Tanganyika (GINDEL 1961).

In all these examples, the concept of adaptation, due to physiological modification in response to environment, is fundamental. Extended experimentation and observation of the phenological, physiological and morphological modifications

15

that occurred during acclimatization of exotic species from 1935 to 1960 (GINDEL, 1956, 1957, 1959), demonstrated that changed environment calls for a response in the plant. The ability to respond, which may be thought of as "plasticity", underlies the process of adaptation and acclimatization.

Acclimatization, as supported by the theory of phytoplasticity, is based on two fundamental concepts: 1. Every plant has the ability to a greater or lesser degree, to change its metabolism in order to conform to a changed environment. The changes in metabolism lead to alterations in the phenology, physiology, and

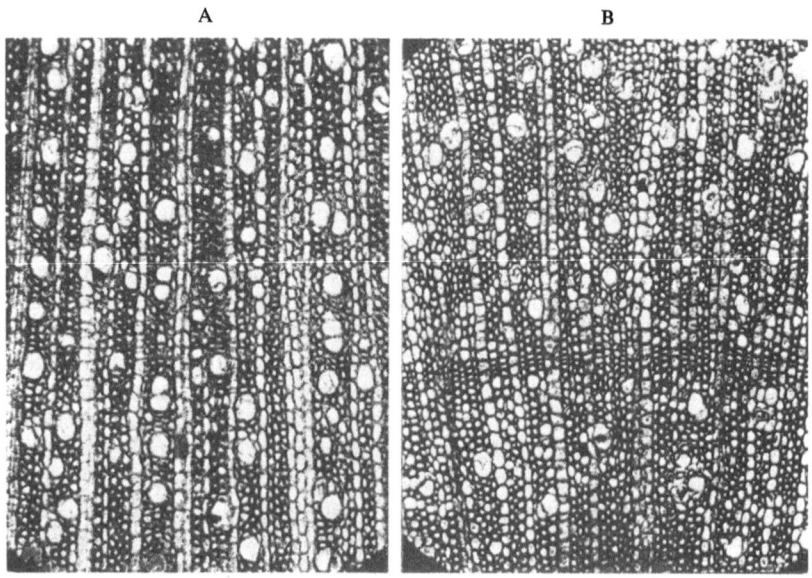

Fig. 7. Coffea arabica L. *var.* Bourbon (Bodr.) Choussy. A transverse section of a four-year old tree (main stem) taken from a plantation at Moshi, Tanganyika (A). A transverse section of the same age and variety, but grown at Rehovot, Israel (B).

structure of the plant. 2. Evolution in the process of acclimatization of the plants is expressed by the increased adaptation of morphological, histological, and anatomical structure, and by the period of physiological activity. The process of evolution involves a continuing preferential selection of individuals with the greatest plasticity, i.e. the greatest ability to adapt.

The 600 woody exotics now acclimatized in Israel originated in latitudes ranging from 60° North to 45° South. On the world scale, exotics from many climates have penetrated into different climatic zones, often changing the original phytogeographic order. Many of the *Eucalyptus* species, numbering at least 630 in their native habitat, are now dispersed in many countries of the world. A tenth of American forest species grow successfully in Europe. Most familiar in Europe, for example, are the olive, orange, nut, grape, and berry trees which are native to

Asia. Wheat is now cultivated from the Equator to 65° North and manifests in all some 400 different morphological and physiological characteristics.

Environment often defines the boundaries of a habitat because of the inability of a particular species to survive beyond those boundaries during the early stages of its development. There are, however, many tree species which, if given the necessary assistance to survive the critical stages, are readily adaptable to the widening of their habitats and extension of their geographical ranges.

The response of plants to changes in climatic or edaphic conditions is clearly evident when the plant is introduced into an entirely different ecological situation. Morphological modifications and physiological changes are sometimes so strikingly apparent that it appears that *the changed environment* may have stimulated a much greater degree of plasticity than the plant had previously demonstrated. When man minimizes the differences between conditions in the natural and in the new habitats of exotics (by irrigating, fertilizing, protecting from sun and wind, etc.), the dynamic reactions of the plant to the new environment are "softened", and the potential plasticity which each plant possesses to a greater or lesser degree is not in evidence. If, however, man does not interfere, critical changes in environment force the plant to summon its total potential of plasticity in order to meet the new demands, and remarkable changes in form and function often result.

Thus it may be said that change in environmental factors challenges or excites plasticity. Within the natural habitat, fluctuations of climate and ranges of plasticity are relatively slight, and modifications or variations within the population are not marked. In a new, often radically different habitat, plasticity is greatly provoked and the plant is able to produce what seem to be radically new responses. For example, the change from mesomorphic to xeromorphic leaves, from homogeneous to more heterogeneous wood, from evergreen to partly or entirely deciduous plants, etc.

The evidence collected by us leads to the conclusion that the ontogenic variations or modifications which an exotic experiences in a new environment will exist indefinitely. However, if the plant is transferred again to a different environment, it will experience other modifications in accordance with the new ecological conditions and will lose whatever characteristics are not essential to the new environment. This explains why many ecotypes of fruit trees, when transplanted by horticulturists in a new environment, fail to manifest the characteristic fruit size or flavor which originally prompted their introduction. For example, oranges of the Israeli mutation, Shamuti, when transplanted in the warm regions of the United States, lose their desired properties.

As illustrated in the beginning of this section, the plant's structure and way of life are influenced by the environmental conditions. Thus, the lack of an ecological approach in studying the plant's structure and physiology often leads to unrealistic conclusions. This fact is reflected in various aspects of the plant, and as an example we shall present the approach in defining the role of the fruit (MAYER, A. M. and POLJAKOFF-MAYBER, 1963), which is in disagreement with its eco-biological properties.

From our experiments (1960), with indigenous and acclimatized fruits,

17

follows that the chemical composition of fruits on aging trees changes to favor seed germination.

The fruit is naturally a disseminating organ, the structure of which is adjusted to the ecological conditions of its habitat.

The teleological notion that flavor and taste of fruit attract the senses of animal and man as a mechanism to disseminate the fruit is not justified. It is true that in nature man or animal consumes the fruit at a stage suitable to his taste. But this is no evidence, at most, that such a mechanism was necessary in the survival of the species.

It should be kept in mind that many cultivated fruits have changed entirely from their original form. This is particularly true of fruits having a pulpy pericarp, which, through selection, hybridization and cultivation, has steadily increased in size, for example, apple, peach, pear, etc. Thus only an ecological approach, a study of the fruit in its natural form and habitat, can elucidate its biological functions.

The developmental cycle of fruits may be divided into four periods:—1. from fertilization until final development in size;—2. the period of chemical changes within the fruit, partly visible to the naked eye by changes in color of the fruit;—3. the period of final fruit ripening during which a corky layer is formed between the pedicel of the fruit and branch of the mother-tree;—4. the period of chemical changes in the fruit when found on its natural litter.

While relatively sufficient information is available on the first three periods, data are scarce with regard to the fourth one. Research has been confined to the period when the fruit is best suited for consumption, or in determining storage conditions most suitable for preserving the fruit. In connection with this problem the following study was carried out (GINDEL, 1960).

Fruits from indigenous indehiscent forest-trees and shrub species were collected soon after they had matured and stored at room temperature until sowing. One hundred fruits from each tree sample were sown in sandy seed-beds.

When testing fruits of *Arbutus andrachne, Ceratonia siliqua* and *Mangifera indica,* the seeds of which germinated in the sown beds, we found that, before sowing, their pH was 4.12, 4.85 and 4.2, respectively. After germination within the fruits their pH increased to 5.38, 6.34 and 5.5 respectively. Even the germination of the seed of the avocado in its fruit is no rare occurrence. In this fruit there is a pH change after harvesting from 6.8 to 7.05 at the time of germination of the seed.

Germination of seeds within the fruit itself was observed in tomato, apple, Artocarpus, integrifolia, Cucurbita maxima, avocado, anona, grapefruit and lemon. Within the last two fruits green cotyledons developed from the germinating seeds when the fleshy pericarp was still entirely closed. The roots of the germs were in all cases tightly attached and had penetrated into the body of the fruit.

From our experiments and observations, it seems therefore clear that the so-called inhibiting substances which prevent the seed from germinating change, in the course of the metabolism of the fruit, into new compounds which favor seed germination. The biochemical composition of the fruit reaches a stage whereby the radicle of the germ is enabled to come in touch with the pulpy content of the fruit and thus provide it with its first food before it reaches the mineral soil in the forest litter.

18

The better results obtained by extracting the seeds artificially should not diminish the positive function of the fruit in Nature as a disseminating organ.

Examining the sown fruits from time to time showed that seed germination occurred wherever the fruit was fresh. On the other hand, wherever germination did not take place, the fruit was found to have been destroyed by fungi or insects. Within the germinating fruit the roots steadily increased in size by making use of the food in the pulpy pericarp, many of these roots branching within the fruits rather than penetrating the soil through the perforated coat.

Are there evidences that fruit is indeed a disseminating organ in the plant kingdom? The anser is yes. It is sufficient to recall the role of fruit in different species

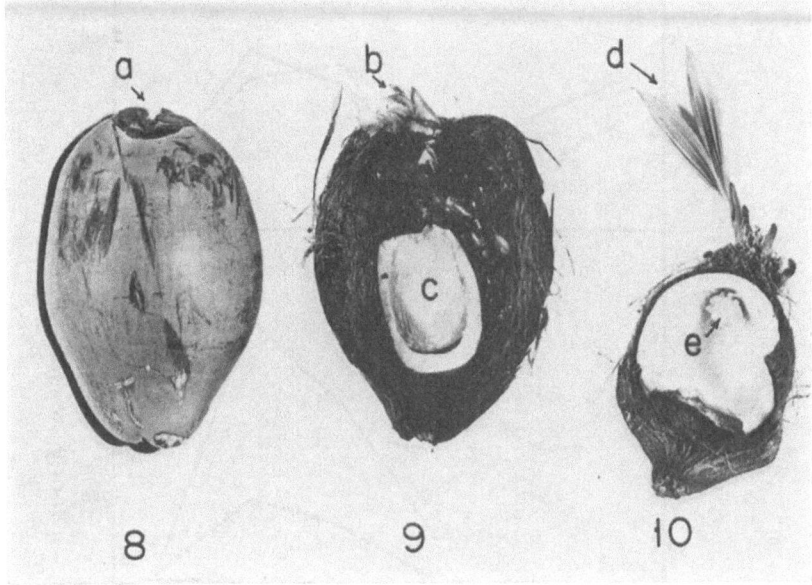

Fig. 8. The eye—a. through which the sprout emerges. *Fig. 9.* Advanced development of the first leaves and roots—b. Within the shell a whitish internal root system absorbing food from the pith-like storage tissue of the fruit c. *Fig. 10.* More developed leaves—d and roots. A fragment of the whitish root system within the "apple" e.

of mangrove forests growing in swamps and along the sea shores (*Avvicennia* sp., *Ceriops* sp., *Rhizophora mucronata, Sconneratia acida* and others).

Under these conditions not only do the seeds germinate inside the fruit which is still attached to the trees, but they also develop into plants 20–30 cm tall. These plants develop a long taproot, and on falling from the tree the plant's root becomes embedded in the swamp. There are numerous examples from tropical countries of many species whose fruits are fleshy, indehiscent and large, such as Artocarpus integrifolia, Cucurbita maxima, Papaya, etc.

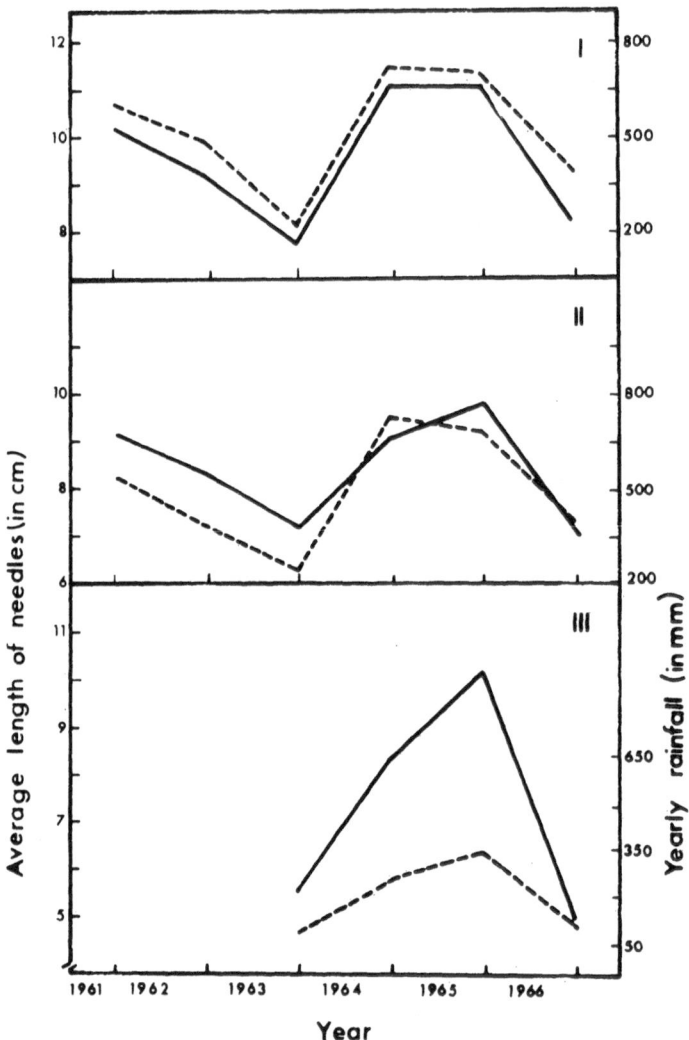

Fig. 11. *Pinus halepensis* Mill. Yearly rainfall (Broken line). Mean needle length (solid line). 1. Mt. Carmel; 2. Judean Hills. 3. Gevulot (Desert).

Another classical example concerning the function of the fruit in the process of germination and on the development of the first leaves and roots is the coconut (Cocos nucifera L.).

After sowing the coconut, the sprout emerges through the eye. When the first leaves sprout through, it starts to develop whitish roots, into the fibrous hull as well as outside the shell. Within the shell, the spongy organ absorbs milk which finally turns into a pitch-like tissue, into which the roots are immersed drawing food for sprouting. Simultaneously 6–8 leaves and apical meristem grow (Figs. 8, 9, 10). This occurs when the first roots did not touch the soil and only the pulp within the shell of the fruit serves as its first germination bed.

The ecological correlation between the environment and the tree can be demonstrated by many other phenomena, as for instance, the clear dependence of mean needle length on yearly rainfall, as illustrated in Fig. 11. The results indicate that an annual rainfall of 200 mm or more in the northwestern part of the desert leads to

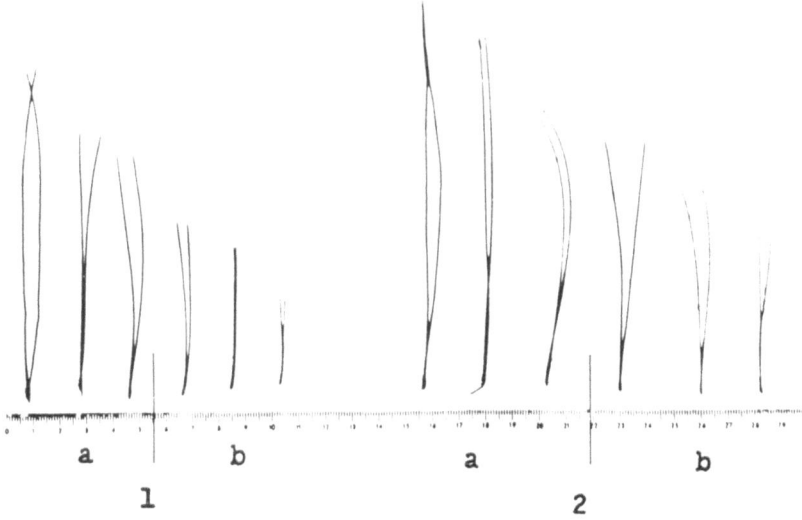

Fig. 12. Mean needle length: I Desert (Gevulot), II Subtropical zone (Rehovot) a. 1964–65, b. 1965–66.

the development of Aleppo pine needles whose length is not much different than those which grow in the Mediterranean zone. When rainfall was only 129 mm during 1965/66 at Gevulot, needle length was at an extreme minimum (Fig. 12).

A similar strong correlation between mean needle length, the annual ring area in the trunk, and annual rainfall was found in two other acclimatized pine species: *Pinus canariensis* and *Pinus pinea*. Groves of these species, together with a grove of Aleppo pine are found in the Judean hills at Kyriat Anavim. All tree groves were planted 28 years ago, forming pure stands. When annual rainfall was 781 mm in 1964, mean needle length was 7.3 cm in Aleppo pine, 9.4 cm in *Pinus pinea* and 23.0 in *Pinus canariensis* (Table I). During 1965/66, when only 502 mm of rain occurred, mean needle length diminished in the three pine species by 20%, 29% and 22% respectively and the annual ring area as follows: *P. halepensis*—36%, *P. pinea*—56% and *P. canariensis*—11%.

21

Table I

The Correlation between Yearly Rainfall, Tree-ring Area, and Average Needle Length

Species	Year	Yearly rainfall in mm	Annual-ring area in cm	Average needle length in cm²	Average tree height in m
Pinus halepensis	1964/5	780.7 (71)	21.74	7.32	13.3
	1965/6	502.0 (51)	13.95	5.79	
Pinus pinea	1964/5	780.7 (71)	23.21	9.40	9.3
	1965/6	502.0 (51)	10.26	6.64	
Pinus canariensis	1964/5	780.7 (71)	20.20	22.79	13.83
	1965/6	502.0 (51)	16.02	17.98	

METHODOLOGY IN FOREST RESEARCH

Since forests are exceedingly complex ecosystems, special adaptations of synecological methods in research are required. A tree is a large and complex ecophysiological unit which exhibits differences in morphology and physiology among 1. the upper, middle and lower parts of the foliage, 2. foliage on different sides of the crown, particularly the northern and southern sides in arid climates, 3. the leaves exposed to direct sun radiation and the interior ones grown in shade, 4. leaves with different orientations, e.g., in Eucalyptus species, ranging from horizontal to vertical, and 5. neighboring stomata, especially when they are present on both sides of the leaf. Research methods must be adapted to the dynamic phenomena of life, and to the complex of ecological factors interrelating and interacting among them.

A postmortem examination of plants, after grinding their tissues, may be more suitable for biochemical studies than for investigating ecological phenomena. It is analogous to a study of human beings from the shadow of a man. The study of ecophysiological phenomena in an environment where some factors are controlled and others are static, may often lead to erroneous conclusions. Often the conclusions do not agree with parallel field observations carried out *in vivo*. We therefore tried to study problems as far as possible outdoors. Even in controlled experiments, the plants were exposed to the natural dynamic changes associated with day and night.

Seedlings of various broad-leaved species kept under glass in order to study water stress dropped their leaves when exposed outdoors during the summer months, and a new foliage developed. The ontogenesis and morphology of leaves of open-grown and glasshouse species differed strikingly.

Examination of forest trees under artificial conditions, e.g., in containers protected by glass or plastic, and irrigated and manured, may lead to erroneous conclusions. The natural relationship of climatic and edaphic factors is radically disturbed, and the ecological laws and the natural correlation of plant and environment are obscured.

Metabolic changes occur when the plant is protected from visible light by means of glass, and from ultraviolet light—even though it comprises only 2% of the sun's radiation. The ultraviolet rays aid in the formation of anthocyanins, and take part in phototropic processes.

Irrigation has far-reaching effects on the structure of the plant and its metabolic processes. It increases the amount of foliage relative to the roots, which in turn leads to heightened physiological activities. Both phenological and morphological changes were found in leaves of irrigated plants. The absorption of salts and the production of hydrophilic colloids are increased by irrigation, as is the amount of free versus bound water. Carbohydrates are rendered as much as $2\frac{1}{2}$ times more mobile.

Fertilizers have an effect on the mechanical properties of the plant, and decrease the amount of oxygen while increasing the proportion of carbon dioxide. In desert trees, which normally have a high proportion of oils, tannins, rubber, etc., irrigation decreased the percentage of combined water. This phenomenon was accompanied by a drop in the osmotic pressure of the cell fluid.

As it will be shown in the course of this investigation, trees in their natural habitats do not undergo conspicuous dehydration even in the desert. This fact is disregarded in studies of the drought resistance of trees, in drought chambers, in order to investigate their limits of resistance to dehydration.

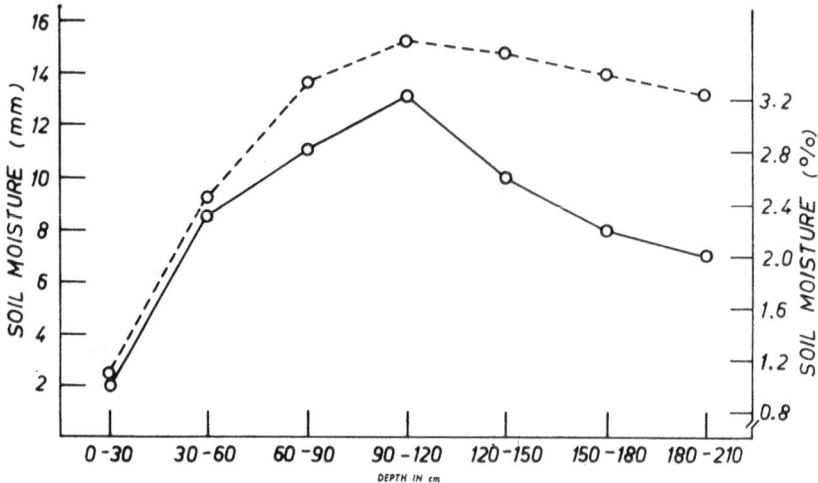

Fig. 13. The difference in fluctuations in soil moisture measured between the gravitational method (solid line in %) and the neutron apparatus (broken line in mm). *Tamarix aphylla* forest. Revivim.

The gravimetric method for studying soil moisture percentage within the forest appeared to yield more precise data than the neutron scatter method, in the studies reported here. Simultaneous measurements by both methods carried out under the canopy of a 14-year old forest planting of tamarisk, in sandy soil, revealed conspicuous differences between the two methods (Fig. 13). It appeared that the neutron scatter readings were influenced by the organic matter and moisture in the roots. Figure 14 shows differences in moisture percentage between the roots and the soil in a forest planting of Aleppo pine in the Hills of Judea.

The necessity of adapting ecological methods for arid zone forestry research is particularly evident in studying the absorption of atmospheric water. Atmospheric water is of decisive importance in maintaining cell turgor in the leaves and the roots of desert plants during long periods without rain. Yet some researchers point out the scarce quantity of atmospheric water absorbed by plants, and its lack of practical value in the water balance (STONE et al., 1950; MAYER and ANDERSON, 1956; WEISEL, 1958). Naturally it is impossible to detect the real quantities of dew and mist that are absorbed by desert plants when hygrophytic or mesophytic plants are studied. Such plants are not adjusted to absorb significant quantities of atmospheric water.

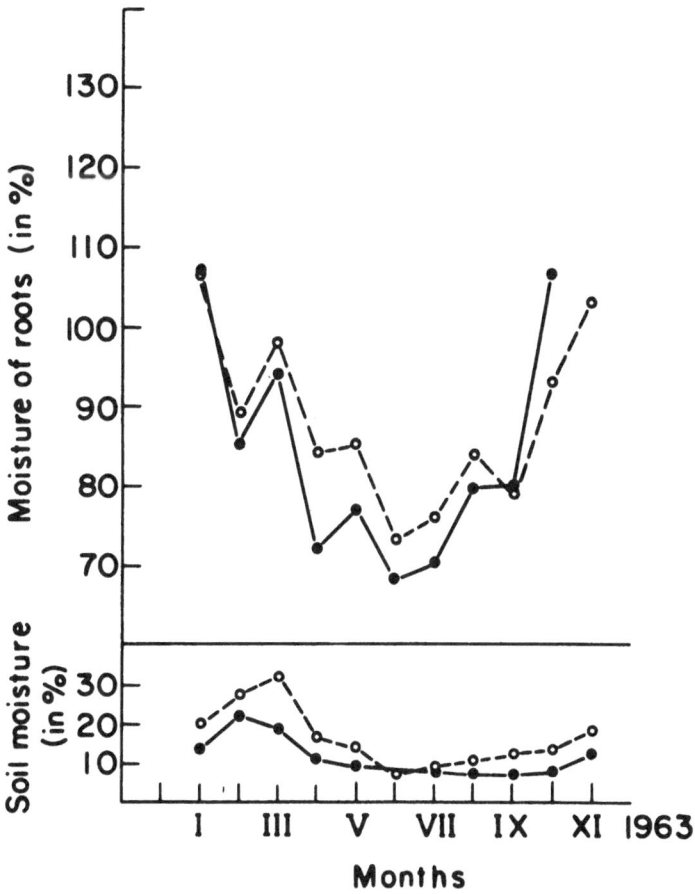

Fig. 14. The fluctuations in soil moisture in Aleppo pine at 2 locations in the Judean Hills, Panorama (broken line), Neve-Elan (solid line). According to LESHEM (1968).

The method adopted in this work to prepare plants for studies in controlled conditions, was: 1. to grow the plants outdoors, i.e , beginning with the germination of the seeds, exposing the plants to the natural climatic factors, 2. to perform the experiments during the hottest summer months when the climatic stress is greatest, and 3. to irrigate sparingly and to keep the plants in water-deficient state. In such a treatment xeromorphic plants are formed. A high degree of soil-water stress and a diffusion gradient are created. The D.P.D. in the air is then less than that in the plant tissues (SLATYER, 1966), and the D.G. of the roots is less than that of the soil.

In order to ascertain the quantity of dew and mist falling nightly in different parts of the country and during different seasons, the apparatus illustrated in Fig. 15 was assembled. It was based on a calculation of the amount of water from dew and mist as a function of area. The apparatus was fitted up in the form of an inverted V with the ratio of height to width—0.5:1.0. Each face was 5 m² in area and was covered with a sheet of polyethylene, 0.1 mm thick. Beside the shorter posts, at the edge of the gutter, were 2 polyethylene bottles into which the water was led through a pipe from a gutter. This "roof" was supported on two pairs of posts of differing height, in the range of 40–60 cm, so as to form a slope which would lead the water into the bottles from the gutters. Polyethylene was chosen as an inactive water collector as it is non-hygroscopic and becomes cool at night. The flow into the bottles began when the sheet was sufficiently moistened, this preventing loss by evaporation. Apart from this means of measuring the water collected on the upper surface of the sheet, reversal of the "roof" enabled the measurement of that forming on the underside, thus giving an overall estimate of the dew.

For irrigation of plants, polyethylene sheets, 1.0–2.0 m² in area, were arranged at a slope of 25–30% (Fig. 16) to facilitate flowing of the water and its transportation to the plant-pit. Plastic ducts were attached to the lower ends of the sheets. To create the necessary slope the soil was ridged or, alternatively, the sheets were placed on wooden boards.

According to CATTELAPPER (1959), indirect methods (infiltration and porometers) are unreliable in studies concerning stomata opening and closing. I have stated this many times. On the other hand, ZELITCH's (1961), method gave precise results.

The cut-leaf, quick weighing method (HUBER, 1927) under field conditions was used in calculating transpiration. It consists of rapid weighing of picked leaves by means of a torsion balance. According to HUBER (1927), the leaf continues to give off water at the same rate as before picking from the tree, and physiological conditions remain unchanged. Successive weighing at 30 sec intervals of detached needles of Aleppo pine to determine the amount of water given off per unit of fresh weight disclosed that there were no fluctuations of the water loss rate until the 10th minute after picking. This method appeared to be useful for comparative studies (RUTTER, 1968). PISEK and TRANGUILINI (1951) and PARKER (1957) have used the method with apparent success. STARK (1967) did not find significant differences when measuring transpiration with Went's transpirometer in comparison to the cut-leaf, quick weighing method.

The method used by us for the last 12 years appears to be quite precise for purposes of comparison. A group of six dominant representative trees grown to full density was selected by a random method in each experimental plot. From each tree, every hour, 10 two-needle clusters were cut and weighed, then weighed again after five minutes. Such a high number of needles, and a relatively long period between weighings, were used to detect very small differences in the transpiration rate between one tree and another, and between one day and another when the transpiration rate was very low, particularly during the hottest and driest months, or during the coldest months of the rainy season (December–February). In transpiration studies additional details should be considered. As the tree grows and develops, some parts show more markedly than other the effect of changes in the hydro-photo-thermal constellation on transpiration.

The rate of transpiration is affected by: 1. Differences in the intensity of light on neighboring leaves; 2. The health of the leaf (damage caused by pests and fungi reduce the rate of transpiration); 3. Clear and cloudy days; 4. Differences in height of the crown; 5. The form of the plant (tree or bush); 6. Site quality. PISEK and CARTELLIERI (1932), for example, showed that in Palo Alto, California, Manzanite and Chamise transpire at the rate of 1.7–2.7 ccs/dec^2 during 24 hours, when grown in chaparral, but only 1.0–1.4 ccs in the forest; and 7. Age of trees. In beech forests, 35, 55, and 115 years old, in temperate climate, rates of transpiration of 2", 9", and 15" per year, respectively, were measured (ZON, 1927). According to GENKEL, et al. (1953) the amount of transpiration is 2.5–8 times greater in young than in old trees. Likewise, the rate of transpiration differs in young and old leaves. Trees at the edge of a plantation in temperate or cold climates show 20–60% more transpiration than those in the center (DVORETSKAYA, 1949), while those at the north side also show a lower rate of transpiration. The following method was adopted to study the transpiration rate as a function of solar energy and soil moisture.

Homogeneously developed seedlings from dominant trees were selected to be planted in small plastic pots, 25 cm^3 in volume. After they had been grown into seedlings 15 cm high they were transplanted into large plastic pots, white in color, 22.5 cm high and 20 cm in girth. Homogeneity of seedlings was preserved to eliminate the influence of differences in growth potential.

In the small and large pots loess-like soil indigenous to the northern part of the desert was used. The large pots were filled with 10 kg of dry soil and within it three plastic P.V.C. pipes, 30 cm long and 6 mm thick were placed into the soil to get the irrigation water into the deeper soil layers. For 21 months a stable soil moisture percentage at field capacity (15.5% by weight) was preserved.

Global radiation, which includes direct and diffuse radiation falling on a horizontal area, was measured with the Epply Pyranometer. The length of measured waves ranged between 0.3–4.5 microns.

Leaf moisture percentage was calculated on a dry-weight basis, after drying the leaves for 24 hours at 100°C. To avoid misleading results, 1-year old full-sized leaves were chosen in accordance with ACKLEY's findings (1954), that in young leaves the amount of dry matter changes during growth.

CHALK's method (1930), was used for following cambial activity. The samples were taken from six dominant trees, in each plot. The wood-samples were taken regularly every fortnight during the years 1945/47, 1954, 1962/63. During the other years between 1945/68, cambial activity was examined during the seasons of interest. They were collected from the main trunk starting at the height of 1 meter with separation between the samples containing the phloem cambium and xylem of the previous two years. After removal from the trunk, they were immersed in a

Fig. 15. Measurements of the water from dew and mist as a function of area. The apparatus is covered with polyethylene.

Fig. 16. Irrigation of pine, tamarisk and eucalyptus seedlings with atmospheric water. At the edge of each polyethylene sheet is a gutter fitted with a small plastic pipe which leads the water into the plant-pit.

glycerine-alcohol mixture (60:40) to facilitate the presentation of 15–20 micron slices by the sliding microtome. Borings were also taken from the trees by means of a Pressler borer and the thickness of their annual rings was measured by means of an instrument developed by the author to an accuracy of 1/10 mm (Fig. 16.1).

During 1962/64, investigation of the cambium was accompanied by measurement of the transpiration rate.

For the establishment of the growth-transpiration relationship, the area, and not the width of the annual tree-ring, was measured, since the 3-dimensional growth of the trunk is more accurately reflected in the area of the ring rather than in its width, which may be the same for a thin, or for a thick trunk.

The density of the stomata and their size (GINDEL, 1969), were compared in irrigated and nonirrigated evergreen and deciduous species, including some fruit-bearing trees. All the species grew in soil where the subterranean water is remote from the root zone. Although two plantations of carob trees (Ceratonia siliqua L.) were both nonirrigated, one was considered irrigated since it grew at the foot of a hill, and consequently the tree roots were able to penetrate into the deep soil of the adjacent valley and thus absorb more water than the second plantation, which grew on the rocky hill ridge. In the case of *Citrus*, both groves were irrigated, but one, which has been included in Table 26 as "nonirrigated", only every 40 days, while the second grove was irrigated every fortnight.

Twenty mature, 1-year-old leaves, which sprouted at the same time, were picked from 20 trees of each category, from the southern side of the crown and from dominant trees fully exposed to the extremes of the climate. Ten spots were measured on each side of the leaf, at the middle of the blade near the midrib, at a magnification of ×450. Two methods were employed to study stomatal density. —1. ZELITCH's (1961) method for measuring stomatal aperture was used soon after picking the leaves, while the cells were still turgid. This consists of putting silica gel (Elastomer R.T.V. III, Rhône-Poulenc, Paris), on the leaf surface. A catalyst to accelerate hardening of the gel (Silicure T-773) was mixed with the gel before applying it on the upper and lower epidermis at the center of the blade. After polymerization, the gel could be detached from the leaf surface. To obtain a positive, the surface of the silicon was smeared with a thin layer of cellulose acetate dissolved in water-free ether.—2. A second method was used for some indigenous species in which leaves are leathery and covered by a thick layer of cutin. The leaves were boiled in O.S.N. NaoH until soft enough to remove the transparent epidermis. On immersing the epidermis in fuchsin, a light-colored dye, the stomata can be seen to protrude clearly from among the epidermis cells.

Seventy-eight tree and shrub species were investigated during the hot rainless months (June–September) between 1966 and 1968, and a comparison was made of their behavior during various hours of night and day (GINDEL, 1970). The heterogeneous group of vegetation included forest and fruit trees, shrubs and green species, and endemic and exotic plants.

Mature leaves were selected on the southern side of the crowns of dominant trees of all species. Silicon rubber impressions were made on stomata while the leaves were still attached to the trees (GINDEL, 1968, 1969a,b) using ZELITCH's technique (1961). The silicon rubber was applied at night, between 11:00 p.m.

Fig. 16.1. An apparatus for measuring tree ring width: 1. Core. 2. Groove. 3. Ocular. 4. Worm gear. 5. Micrometer dial. 6. Scale. 7. Micrometer.

and 3:00 a.m., when dew condensation on the leaf surface was apparent. In species possessing stomata on both sides of the leaf, silicon impressions were made simultaneously on both surfaces. Stomatal aperture was measured with a planimeter using enlargements of photomicrographs.

The seedlings were grown in pots with a controlled soil moisture content, and their exposure to the extreme climatic conditions out-of-doors resulted in the development of a xeromorphic leaf structure with the increased ability to absorb atmospheric moisture.

THE CONSUMPTION OF SOIL WATER BY TREES

Two categories of woody plants may be distinguished when tree-water relationships are analyzed: 1. Indigenous or domesticated tree species grown naturally. 2. Those grown with man's assistance (irrigation, manuring, sheltering from wind, and shading). In the two cases the problem of water requirement and its consumption is different. In fact, little similarity can be found in regard to plant water balance between forest and horti- or agricultural-irrigated plants. In this case the results of investigations carried out with indigenous and exotic-domesticated trees grown without interference of man will be analyzed.

The main problems in regard to water fluctuations within the root zone in extreme climatic and edaphic conditions, are as follows: How deep does rainwater penetrate into the soil? How much available water is left for the tree after gravitational water has drained beyond the root zone during the rainy season? How much water is evaporated from the soil until and during the season when the roots start to absorb the soil moisture? And what is the effect of years of drought on soil moisture depletion?

Topography and the physical properties of the soil often are decisive insofar as infiltration is concerned. The maximal penetration was noted in sandy soil, and the lowest on sloping land with thin soils. In the hills, characterized by a thin layer of soil above the mother rock, only a small part of the rainwater percolates into the soil. This is particularly true on steep slopes and when the rains are intense, a characteristic feature of the Mediterranean climate.

Figure 18 shows the result of soil moisture measurements on a northern-exposed 30% slope. At the end of the rainy season (March), an average moisture content of 26.3% for the three 30 cm layers was observed in the bare area. Figure 20 shows the water penetration into a sandy soil under the canopy of a Tamarix planting in the desert where 104 mm of rain occurred. Moisture measurements were taken on 5/5/61. In bare areas a conspicuous quantity of rainwater had already evaporated. Under the canopy of the plantings presented in this work the highest percolation reached a depth of 60–120 cm. In April 1963/64 when yearly rainfall was 165 mm, the maximum percentage of rainwater in the bare area was found at a depth of 90–120 cm. In 1964/65, an exceptionally rainy year, (310 mm) the maximum percentage was at a depth of 240–270 cm (Fig. 21).

In order to compare the absorption and depletion of soil moisture in artificial and natural forests, simultaneous studies of adjacent bare areas of similar topography and soil structure were carried out. These artificial and natural forests grow in dry rocky soil in the hills, or on sterile deep soils in the desert and semi-desert, where the root system is far from underground water. In the hills a soil layer of only 15–20 cm covers the bed rock. In the desert the soil was tested to a depth of 240–330 cm (GINDEL, 1961, 1964, 1967).

A comparison of soil moisture percentages under the canopy of forest and in treeless areas, as well as their seasonal fluctuations, is one way of determining the amount of water used by the forest. In the open areas, soil water is evaporated by

Table II. Soil Moisture within and Outside the Forest (Judean Hills)

Date of Testing	Species	Soil Type	Depth (cm)	Within the Forest %	Within the Forest mm	Outside the Forest %	Outside the Forest mm	SE[1]	N[2]
Oct. 1960	*Pinus halepensis*	Terra rossa	0–30	9.5	35.6	6.2	23.2	0.65	20
			30–60	13.9	54.3	9.3	36.3	—	—
Nov. 9, 1961	*Pinus halepensis*	Terra rossa	0–30	9.6	36.0	7.5	28.1	0.82	21
			30–60	11.3	44.1	10.7	41.8	—	—
Nov. 9, 1961	*Pinus halepensis*	Rend-zina	0–30	7.2	27.0	6.3	23.6	0.82	21
			30–60	10.1	39.4	9.7	32.8	—	—

[1] S.E. = Standard Error.
[2] N = Number of samples.

seasonal vegetation, direct sun-illumination, high temperatures, and strong winds (often dry and hot) which play their part in exhausting soil moisture. The soil moisture percentages inside and outside the forest were analyzed statistically to determine if the differences were indeed significant.

Table III

Soil Moisture within the Forest (I) and in the Open Area II, Judean Hills, October 1961

Species	Location	Depth of soil layer (cm)	I %	I mm	II %	II mm	S.E.	N
Pinus halepensis	Ma'aleh	0–28	12.8	48.0	7.9	29.6	0,60	20
Pinus halepensis	Eshtaol	0–32	11.1	41.6	7.9	29.6	0.85	20
Pinus brutia	Eshtaol	0–28	10.6	39.7	7.1	26.6	0.88	20
Eucalyptus gomphocephala	Eshtaol	0–33	15.3	57.4	10.1	37.8	0.65	20

Within the forest, soil moisture was tested under dominant trees belonging to the main canopy. Every soil sample boring in the forest was accompanied by a similar boring in the open area at a distance af 50–100 meters. The open areas were covered every year with seasonal vegetation, most of which sprouted after the first rains, and died by the end of May, after which the foliage turned yellow.

34

Fig. 17. Pinus halepensis, 27 years old. Planted in soft limestone on the hills of Judea, Superficial and horizontal ramification of root system 30–50 cm deep; only one root penetrates into a horizontal soil pocket 85 cm deep.

The differences in edaphic moisture and its variations with depth were examined under the following species: *Pinus halepensis, P. brutia, Schinus molle, Cupressus pyramidalis, C. horizontalis*, and the indigenous, woody vegetation in the hills belonging to the Mediterranean maqui. In the desert *Tamarix aphylla, T. pseudo-pallassii*, and various acclimatized species, such as *Eucalyptus gomphocephala, E. camaldulensis, E. occidentalis*, and *Acacia cyanophylla* were studied. Also, soil samples were taken beneath the foliage of indigenous woody plants found in the desert, such as: *Acacia tortilis, Retama roetam, Haloxylon persicum, Thymelaea hirsuta*, and others.

The Aleppo pine was the main forest tree investigated in the subtropical region. It has marked xerophytic characteristics and is one of the few pine species growing naturally on chalk formation; however, it also grows well when planted on the terra rossa soils. Its root system is superficial and is mainly confined to the thin soil layer. The majority of its roots spread horizontally throughout the superficial layer of soil, often reaching no deeper than those of some natural herbaceous perennials (Fig. 17).

Three consecutive drought years (1956/57–1959/60) created favorable conditions for investigating the degree to which soil moisture was exhausted under the canopy of the forest planting, and in open areas covered by seasonal grasses. For these years the relative effectiveness of rainfall was determined by calculating the aridity index; it was one-third as great as for the three previous years (1953/54–1955/56) when there was normal rainfall (GINDEL, 1966). Differences in the amount of soil moisture, and its pattern at the end of the third drought year, are presented in Tables II, III and IV.

Table II shows the moisture percentage in terra rossa beneath the foliage of a 30-year-old *Pinus halepensis* stand on a 15% slope. In October, 1960, soil moisture to a depth of 60 cm in the forest was found to be from 3.4–4.7% greater than in the open area. At the end of the dry season of the following year (November 9, 1961) at the same locations, similar differences were found in both terra rossa and rendzina soils. Lateral root development was greatest in the upper two layers where the moisture content was greatest.

Another four sets of sample plots were established in two locations along the slopes of the Judean Hills in 1961 (Table III). Here soil moisture percentage during one of the critical months (October) is higher under the canopy of trees than in the open area. The difference is greatest in a Eucalyptus grove, and the lowest in a *P. halepensis* grove. The differences were statistically significant at the 0.01 level in all four plantings.

The observed changes in soil moisture from the beginning to the end of the dry season aroused an interest concerning the monthly rates of water depletion under the canopy of *P. halepensis* plantings as compared to treeless areas. Investigations were carried out in seven sets of sample plots in four regions. In all seven locations, significant differences at the 0.01 level were found between the different months (Table IV). A comparison between forests and open sites in three localities (Mt. Canaan, Nazareth and Hulda), showed that the soil moisture content was significantly higher in the forest in all months (Table V). The excesses ranged from 3 to 30% in terra rossa, and from 8 to 31% in rendzina.

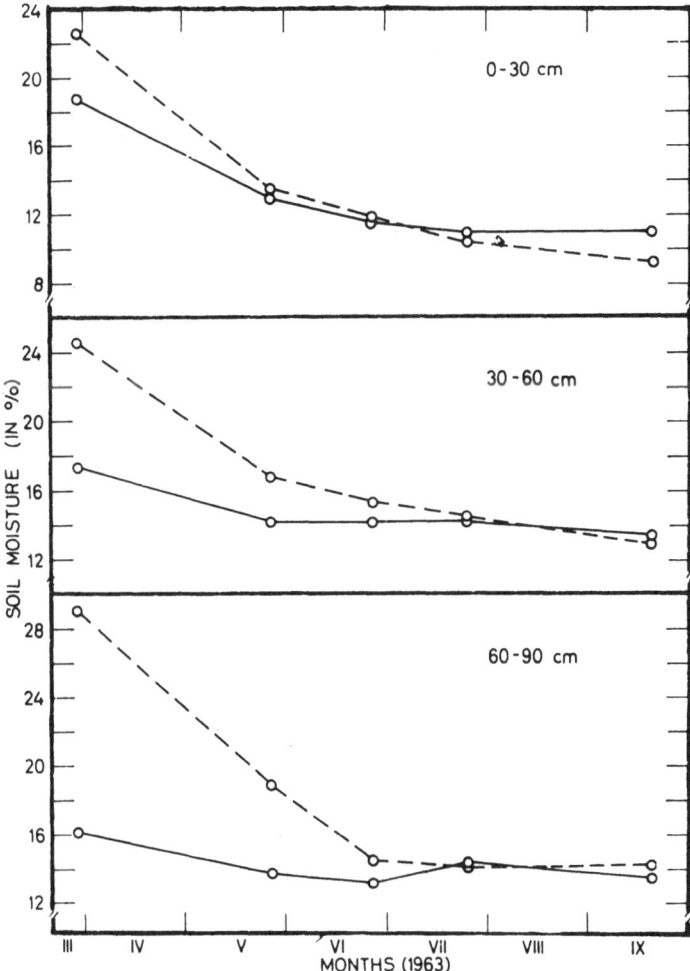

Fig. 18. Soil moisture fluctuations in three soil layers: under the forest canopy (solid line) and in the bare area (broken line). Efraim Hills. Aleppo pine.

In the hills of Efraim, in a 12-year-old Aleppo pine plantation, the seasonal trends in soil moisture at various depths throughout the rainless period are shown in Fig. 18. On March 28, 1963, when the seasonal vegetation was flourishing, just seven days after precipitation of 1.4 mm, the moisture was greater in the treeless area by 4% for the 0–30 cm layer, by 7% at 30–60 cm, and 13% at 60–90 cm. On this date the moisture content in the forest varied between 16 and 18%. By July 6, 1963, the differences between the forest and treeless areas had disappeared. From the beginning of the dry season until May 26, 1963, soil moisture was higher in the

37

Table IV. Monthly Moisture Fluctuations at Seven Locations (1962)*

Location	Soil type	Depth (cm)	Soil Moisture								SD[1]	CV[2]
			%	mm	%	mm	%	mm	%	mm		
Mt. Canaan S.P. 19 and 20			July 2		Aug. 7		Sept. 24					
Within the forest	Terra rossa	0–30	15.7	58.8	14.5	54.4	14.2	59.2			1.34	9
		30–60	—	—	20.5	80.0	19.2	75.0				
In the open area		0–30	12.6	47.2	10.2	38.2	10.4	39.0			1.97	18
		30–60	—	—	19.2	75.0	—	—				
Nazareth Hills S.P. 21 and 22			June 12		Aug. 7		Sept. 10					
Within the forest	Terra rossa	0–30	12.5	46.8	10.3	38.6	10.0	37.5			1.61	15
		30–60	13.9	50.7	12.2	47.6	11.0	41.9				
In the open area		0–30	10.9	40.9	8.0	30.0	8.4	31.4			1.98	22
		30–60	12.2	42.6	11.52	45.3	11.0	42.7				
Mt. Carmel S.P. 23			Apr. 11		May 23		June 21		Sept. 10			
Within the forest	Terra rossa	0–30	15.6	48.5	14.3	53.6	13.1	49.1	10.9	40.9	2.69	20

Judean Hills Eshtaol S.P. 24 — Within the forest — Terra rossa (0–30)

Apr. 16		May 14		June 27		July 27		Sept. 11		SD[1]	CV[2]
17.6	66.0	16.9	63.4	13.1	49.1	13.6	51.0	13.8	51.7	2.15	14

Judean Hills S.P. 25 — In the open area — Terra rossa (0–30)

May 17		June 27		July 27		Sept. 11		SD[1]	CV[2]
11.3	42.4	9.1	34.1	9.5	35.6	8.6	32.3	2.26	25

Judean Hills Hulda S.P. 7 and 8 — Rendzina

	Depth	May 20		June 29		July 31		Sept. 3		SD[1]	CV[2]
Within the forest	0–30	7.8	29.3	6.9	23.0	6.4	—	6.0	22.5	1.14	17
	30–60	—	—	—	—	9.3	—	8.2	32.0		
In the open area	0–30	6.7	25.1	5.0	18.8	5.3	—	4.2	15.8	0.91	17
	30–60	10.5	40.1	9.6	37.5	—	—	7.5	29.3		

Judean Hills S.P. 26 — Within the forest — Rendzina (0–30)

May 17		June 27		July 27		Sept. 1		SD[1]	CV[2]
10.0	37.5	8.9	33.4	6.6	24.8	7.0	26.3	1.94	24

* The non-significant monthly decreases or increases are in italics.

[1] SD = Standard deviation.

[2] CV = Coefficient of variability.

treeless area, and the difference was statistically significant at the 0.01 level. From June 26 to September 17, 1963, the differences between the treeless area and the forests were not significant even at the 0.05 level. However, by September 17, soil

Table V. Moisture Excess within the Forest in Comparison to the Open Areas

Location	Date of testing (1962)	Soil type	Depth (cm)	Moisture excess (%)
Mt. Canaan	July 2	Terra rossa	0–30	19.7
S.P. 21, 20	Aug. 7	Terra rossa	0–30	29.6
	Aug. 7	Terra rossa	30–60	6.4
	Sept. 24	Terra rossa	0–30	26.8
Nazareth Hills	June 12	Terra rossa	0–30	12.5
S.P. 21, 20	Aug. 7	Terra rossa	0–30	22.3
	Aug. 7	Terra rossa	30–60	4.5
	Sept. 10	Terra rossa	0–30	15.8
	Sept. 10	Terra rossa	30–60	3.3
Judean Hills	May 20	Rendzina	0–30	13.4
Hulda. S.P. 7, 8	June 29	Rendzina	0–30	18.1
	July 31	Rendzina	0–30	18.2
	Sept. 3	Rendzina	0–30	30.8
	Sept. 3	Rendzina	30–60	8.2

Table VI. Percentage Moisture within the Pine Grove and in the Adjacent Treeless Area at the Southern Border of the Judean Hills

	Moisture	Content (%)
Soil depth (cm)	Inside the forest	Open area
15	4.0	2.3
30	6.1	3.8

moisture within the forest was higher in the 0–30 cm stratum (significant at the 0.01). A similar situation appeared at the southern limits of the Judean Hills, facing the desert, where rainfall was 244 mm during the year of measurement, soil moisture surplus under the canopy were 43 and 38 % depending on the depth of the soil layer (Table VI).

More surprising soil moisture surpluses was discovered under the canopies of forest plantings in comparison to bare areas in the desert where yearly rainfall is only 100–200 mm and during years of drought only 30–50 mm.

The *Tamarix aphylla* groves, for instance (Fig. 19) were 12–14 years old, planted at distances of 1 × 2 m in sand dunes and in stationary sands, derived from Nubian sandstone and from igneous rocks. About 800–1000 trees per hectare survived. At the time of the observations, the height of the trees was 6–8 m and their average diameter was 12–14 cm.

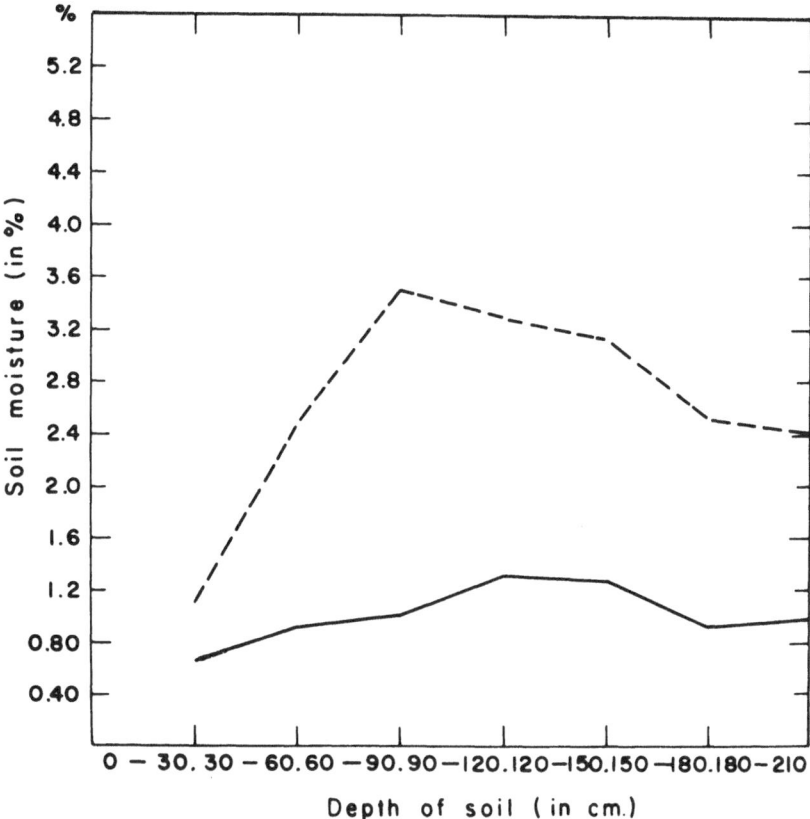

Fig. 19. Percentage of soil moisture beneath the canopy of a sixteen-year-old wood of *Tamarix aphylla* (broken line) and in an open area (solid line) at the end of the dry season. Sand dunes. Revivim.

The same pattern in soil moisture distribution was found the following year (October 21, 1961) in the Revivim forest (*T. aphylla* in sand dunes) after the rainy season (5/5/61, Fig. 20), and before the rain occurred (21/10/61).

Figure 21 illustrates the soil moisture situation in 1964/65 in the Tamarix and Eucalyptus in Gevulot, both 23 years old, and in a nearby treeless area. Rainwater did not penetrate deeper than 330 cm. The maximum percentage of moisture in the

Tamarix grove was observed at a depth of 1.50–1.80 m. On November 1, 1964, edaphic moisture was greater in the Tamarix grove than in the open.

Despite a heavy rainfall of 309 mm during 1964/65, the maximum percentage moisture under the canopy of *Eucalyptus* was found at a depth of 240 cm inside the grove, and at 270 cm in the exposed area.

Table XXVIII, shows soil moisture percentages under the canopy of a *Tamarix aphylla* plantation, and open area adjacent to it at the end of the dry season in four different years. The average percentage is 2.25. The differences among the years are

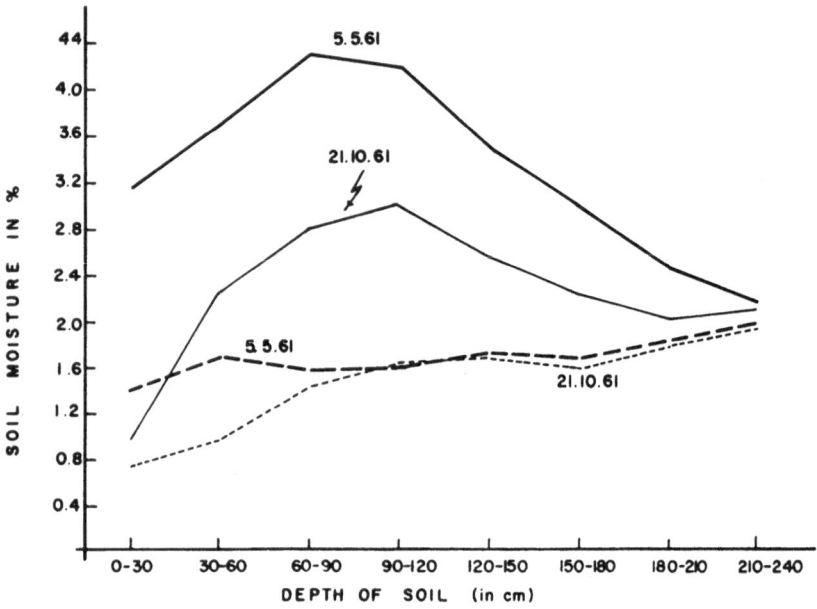

Fig. 20. Differences in soil moisture beneath the canopy of *Tamarix aphylla* (solid line) and in an open area (open line) before the rains and after them. Sand dunes. Revivim. 20.10.1961.

not significant. The average in the bare area is 0.96%. It follows that the quantity of soil moisture under the canopy of Tamarix at the end of the dry season is independent of the amount of yearly rainfall (between 44 and 165 mm), and is consistently somewhat higher than in the bare area.

The soil water balances after three consecutive drought years show how much moisture the indigenous and acclimatized xerophytes preserved during the driest and hottest months. For example, the 559 m³ per ha of moisture found under Tamarix in October, 1960 (before the rains fell) was sufficient to cause it to flower.

The forest plantings mentioned above were established on an area which had been treeless for thousands of years. Before planting, the afforested sample plots

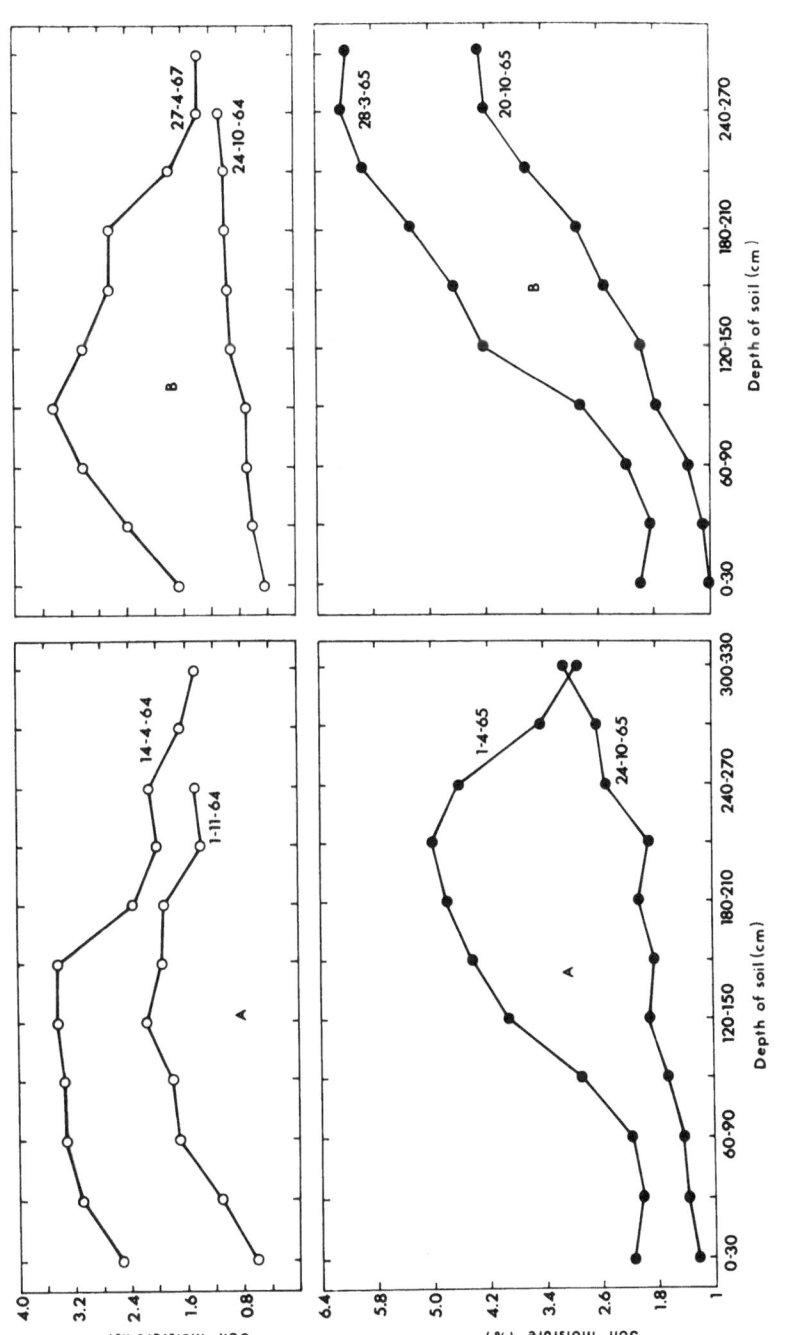

Fig. 21. A profile of the edaphic moisture at the beginning and end of the dry season within the grove—A. and in the exposed area—B.; top: a *Tamarix* grove; bottom: *Eucalyptus* grove. Gevulot.

looked like the present bare areas. Of course, during the 12–14 years of growth in the desert and in the subtropical zone, there has been an addition of a 5–8 cm of forest litter. But it has not yet been able to decompose and to influence the bulk density of the soil.

Higher soil moisture was observed not only in plantations, but also under naturally grown indigenous trees and shrubs belonging to the Mediterranean maqui. The excess was found even under isolated evergreen trees. The shade of the tree canopies is not the reason for higher moisture in the soil. No significant moisture decrease was found under the canopy of *Pinus halepensis* from the end of June or July, 1962, until the onset of rainfall October 1962. In all three climatic regions of the country, more moisture was found during the rainless season under the canopies of forests than in the open treeless areas.

In the open areas, soil moisture shows an entirely different pattern, often slightly increasing with depth. There the soil moisture distribution results not from the action of tree roots, and from the absorption of atmospheric moisture, but from the action of seasonal vegetation, temperature, wind, and prolonged direct illumination. The direct action of thermodynamic factors is not confined to the superficial soil layers, but evaporates moisture to a depth of 2–3 meters in the desert, in sandy soils. At a depth of 240–330 cm, depending on location, species, age of planting, and quantity of rainfall, moisture percentage is equal outside and inside the forest at all times.

The changes found under the forest canopy are due to the action of the planted trees only, particularly in respect to the differences in moisture content and its distribution. By testing the structure and distribution of the roots of the various tree species, it appears that where underground water is far from the roots and

Table VII. Fluctuations in Annual Rainfall and in the Percent of the Soil Moisture in the Forest during 1961/2 (Normal Amount of Rainfall) and 1962/3 (a Very Dry Year).

Species	Location	Rainfall in mm	Date of testing	Depth of soil in cm	Soil moisture %	Soil
Pinus halepensis	Mt. Canaan	1961/2	7.8.62	0–30	14.5	Terra
		756(45)		30–60	20.5	rossa
		1962/3	15.8.63	0–30	14.3	
		601.6(41)		30–60	20.0	
Pinus halepensis	Eshtaol	1961/2	11.9.62	0–30	13.8	Terra
		486(42)				rossa
		1962/3	29.9.63	0–30	12.0	
		222.5(20)				
Pinus halepensis	Hulda	1961/2	3.9.62	0–30	6.0	Rendzina
		407.0		30–60	8.2	
		(42, 12)				
		1962/3	29.9.63	0–30	7.0	
		238.5(21)		30–60	9.1	

where rainwater and atmospheric moisture are the only sources of water, the root structure is superficial, branching laterally and diagonally.

The influence of the physiological activity of trees on the water balance is striking when soil moisture in years of normal rainfall is compared to years of drought. Table VII compares the summer soil moisture in 1961/62 (a normal year) with 1962/63 (a year of drought). On Mt. Canaan, the rainfall was 20% less in 1962/63 than in 1961/62, yet there was no significant difference in soil moisture in the forest. In Eshtaol, on the same type of soil (terra rossa), the rainfall dropped to 46% of normal in the drier year, while the soil moisture decreased by only 9%. In Hulda it dropped to 61%, while the percentage moisture in the two upper layers increased by 11%.

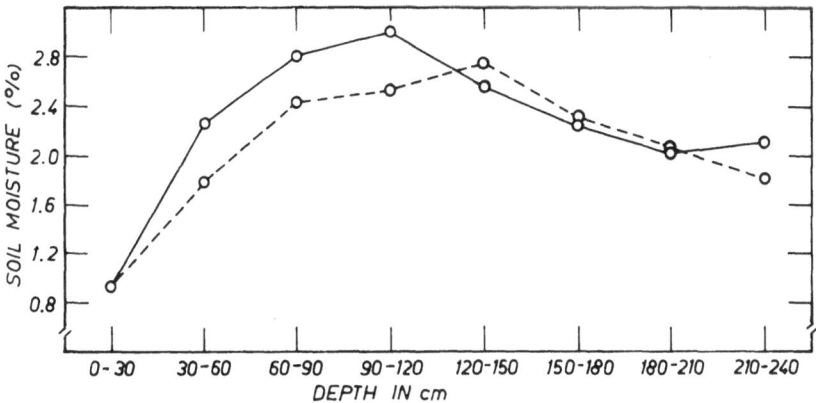

Fig. 22. *Tamarix aphylla* forest. Soil moisture fluctuations in 1961/2—Rainy year (solid line and in 1962/3—Dry year (broken line). Sand dunes, Revivim.

In 1961/62, in the Tamarix forest at Revivim, it rained 104 mm during 32 rainy days, while in 1962/63, there were only 8 rainy days and a total of 32 mm. Figure 22 shows that there was no marked difference in soil moisture for these two years. The mean decrease in the dry year was only 0.5% in the 60–130 cm stratum. As shown by statistical analysis, no significant differences were observed among individual strata from 30–150 cm. (The S.E. varied between 0.1 and 0.2.) In contrast, a very significant difference (at the 0.01 level) was measured in a neighboring exposed area (Table VIII). The difference was almost 100% at a depth of 60–90 cm. The S.D. (0.17–0.16) increased with depth.

The accepted equation for calculating evapotranspiration agreed with the results of soil water depletion in arid regions.

$$ET(mm/day) = \frac{M}{100} \times BD \times L$$

Where M = soil moisture difference, in percent by weight, daily mean for the dry season, for a given stratum, BD = bulk density of the soil in the stratum, and L = depth of soil layer in mm.

It follows from Table IX that the daily evapotranspiration in the treeless areas exceeded that in the forests. The actual evapotranspiration is shown for two

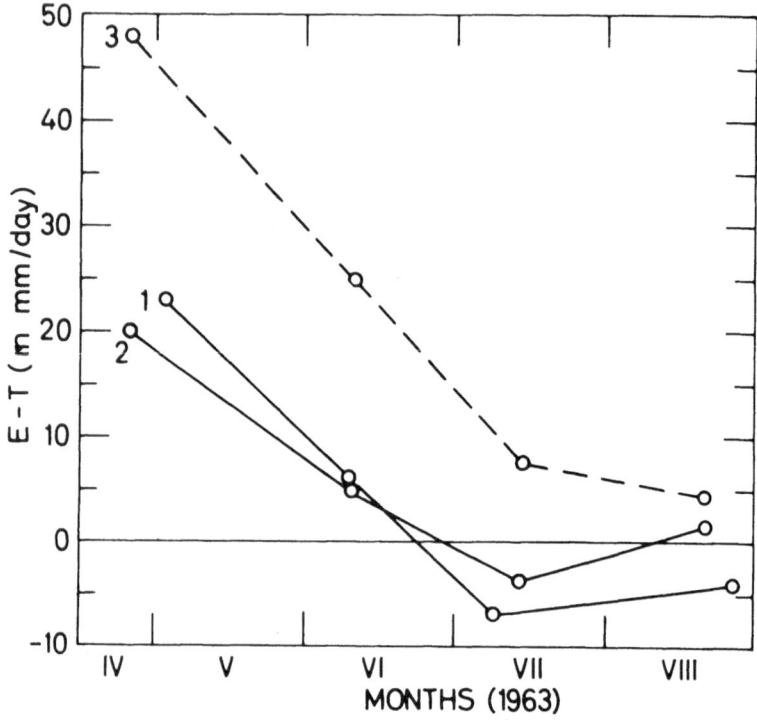

Fig. 23. Seasonal trends in daily evapotranspiration (E–T) under the canopy of Aleppo pine (1, 2), and in the treeless area (3). Efraim Hills (1, 3) and Judean Hills (2).

Table VII. Soil Moisture in Exposed Dunes at Revivim in 1961 and 1963

| Depth of layer | 21/10/61 | | 13/10/63 | | S.D. |
	%	mm	%	mm	
0– 30	0.8	3.8	0.5	2.4	0.2
30– 60	1.0	4.8	0.6	2.9	0.4
60– 90	1.4	6.7	0.8	3.8	0.5
90–120	1.6	7.7	1.1	5.3	0.6

Table IX. Evapotranspiration during the Rainless Season in Treeless Areas and in Forest Plantings of Aleppo Pine and Tamarisk

Sample plot No.	Species	Location	Elevation a.s.l. in meters	Exposition	Soil	Depth in cm	Dates of soil moisture tests at beginning of dry season	Dates of soil moisture tests at end of dry season	Evapotranspiration in mm/day		
									Treeless area	Forest canopy	Difference in %
1	Aleppo pine	Mt. Canaan	250	West	Terra rossa	0–30	2.7.62	24.9.62	8.0	5.4	32.5
2	Aleppo pine	Judean Hills	250	South	Rendzina	0–30	20.5.62	3.9.62	6.8	4.9	37.9
3	Aleppo pine	Judean Hills	350	South-west	Rendzina	0–30	14.5.72	11.9.62	12.2	7.8	36.1
4	Aleppo pine	Efraim Hills	310	North	Rendzina	0–90	28.3.63	19.9.63	22.9	15.6	36.5
5	Aleppo pine	Judean Hills	360	North	Terra rossa	0–90	4.4.63	24.9.63	—	13.1	—
6	Tamarisk	Revivim	280	—	Sand dunes	0–270	28.3.64	19.9.64	4.1	3.23	21.3

experimental plots (Fig. 23). Within the forest it is similar for both plots, even though they are located in different soil types (rendzina and terra rossa) and in different ecological regions. In the treeless area, evapotranspiration continues throughout the summer months until the beginning of the rainy season; it is much greater than inside the forest. Because of this, the excess moisture in the soil, compared to that in the forest, declines and finally disappears during the summer. Beginning in June, there is more moisture in the forest soil, and this state of affairs continues until the rainy season arrives.

These results demonstrate the impracticability of using physical equations to imitate physiological phenomena in nature. Such equations usually postulate that soil moisture decreases with proximity to the sphere of influence of the roots. Within the subtropical and desert areas under review, studies have shown the opposite state of affairs to prevail after the available soil moisture has been absorbed by the roots. During the driest season, among xerophytes grown without irrigation, the moisture content of the soil increases near the roots. Altogether different definitions must be applied, therefore, to woody xerophytes grown naturally in arid conditions, without the assistance of man.

ABSORPTION OF ATMOSPHERIC MOISTURE BY WOODY XEROPHYTES

From the previous chapter it can be seen that the evergreen xerophytes retain full turgor in their leaves and continue their normal physiological activity until the rainy season begins in November. All this takes place despite the fact that the available soil moisture in the root zone was depleted by the end of June or July. Furthermore, the minimum soil moisture level after this period is greater than that in an open, unforested area identical to the wooded area. This surplus in the root zone is no different in rainy years from in years of drought with only a third or a quarter of the normal rainfall, and it is constant for each species, region and soil type. Since the studies on this subject were carried out under arid soil conditions where there was no ground water, nor any other water reserves reaching the roots, and since only the rainfall was available to the roots, we studied the only source of water which could enable the leaves and roots to retain their full turgor condition during these critical periods—the atmospheric water. Measurements of dew in various parts of the country showed that it increases in quantity from the relatively rainy subtropical north to the south, and in the northwestern part of the Negev the dew fall reaches a yearly maximum of 150 mm (ASHBEL, 1936).

The seven years 1956/57–1962/63, were mainly drought years, a relatively rare occurrence in the history of meteorology in Israel. They created satisfactory conditions for investigating the influence of dew and mist on the water balance of both indigenous and acclimatized xerophytes. These trees continued to grow during the critical summer season without showing any signs of foliage wilting. At the beginning of the dry season of the seventh year, the trees appeared as usual phenologically, even though in this year the rainfall was minimal in both the sub-tropical and desert zones. From the march of the moisture depletion in the root zone of the forest, it was soon seen that after a *status quo* has been reached, the evergreen species which maintain their green foliage during the critical months (July–October) until rainfall starts, avoid desiccation by absorbing atmospheric moisture. Due to the absorption of atmospheric water, no significant differences were noted in soil moisture percentage during the critical months of normal and drought years.

The modern farmer in Israel knows from his daily contact with nature that dew and mist are two factors upon which he can rely. Certain dry crops, such a maize, wheat, barley, and sesame, flourish where there is heavy dew (as in the western Valley of Israel as compared with the eastern section of the Valley, which is less fortunate in this respect). In the desert again, it has been found that grain crops can withstand a dry season in the northwestern desert (rich in dew) better than in the east in which there is little dew.

The ancients, from long personal experience, appreciated the importance of dew for their crops. This is attested by many verses in the Bible. ISAAC in blessing JACOB said: "So God give thee of the dew of heaven, and of the fat places of the earth" (Genesis 27: 28), while MOSES in the first of his valedictory poems to Israel before his death in MOAB juxtaposed "dew" with "rain" ("My doctrine shall drop as the rain, my speech shall distill as the dew"—in Deuteronomy 32: 2), and further, in the next chapter there is an even more significant reference—"For the *precious things of heaven*, for the dew" (33: 13) and again, "Yea, His heavens drop down dew" (ibid. 28). The prophetic books abound in allusions to the beneficent influence of dew for the tiller of the soil and a few examples will suffice. MICAH says, "The remnant of JACOB shall be . . . as dew from the Lord, as showers upon the grass" (5, 6—again potent recognition of the importance of dew, akin to that of rain). "I will be as dew unto Israel; he shall blossom as the lily"—(14: 6); and ZACHARIAH: . . . "the ground shall give her increase and the heavens their dew" (8: 12). Naturally, the withholding of dew is a disaster and the best known reference in such a context is the passage in DAVID's lament over SAUL and JONATHAN: "Ye mountains of Gilboa, Let there be no dew nor rain upon you". (2 Samuel 1: 21).

It is a matter of fact that LEONARDO DA VINCI, whose thoughts embrace many phenomena in plant life, expressed his opinion that the structure of the leaf is adjusted for absorption of dew. On the other hand, the opinion of modern science is split into two parties. While one school of thought belittles the practical effect of dew (SCHRODTER, 1950; MEYER, B. S. and ANDERSON, D. B., 1960; MONTEITH, I. L., 1963), its supporters found more encouraging results (ZAMPIRESCO, 1931; DUVDEVANI, 1947; STONE, 1950; EISENZOPF, 1952).

The "Loma" vegetation in the Atacom Desert of northern Chile and Peru, where rain is negligible and infrequent, lives by means of dew and mist. Although there is only 50 mm rainfall annually, heavy mists occur for six months of the year and provide moisture equal to 300–400 mm of rain. In the Namib Desert in South West Africa, the total depth of annual mists is 200 mm, while rainfall is only 39.7 mm. In the deserts of Nomagualand and Umdons, succulent plants which gather water from the mists, actually appear within the rocks, for example: *Aloe, Euphorleier, Hoodia Lithops, Pelargonum, Soncrecoulen, Trichocoulen* and others.

BLACK (1954) in his research on 2 Atriplex sp. grown in the Australian desert including the results of other workers argues that the vesicular tissue in the above mentioned genus, grown in arid conditions, stores water and is a medium to allow the rapid absorption of atmospheric moisture in the mesophyll.

Absorption of water vapor by the parenchyma of the leaves in three dominant shrubby species, in the Death Valley, California, was noted by Dr. STARK and LOVE (1969); Larrea divaricata, for instance, has a sticky mucilaginous material on the leaves which appears to be hygroscopic.

The more pin-like the leaves are, the bigger is the chance of a condensing of small droplets on them during night time. This can take place even if the air seems dry.

The Cactaceae present a special biological unit; in spite of growing often in an extremely arid habitat, and in rocks, their osmotic pressure is not higher than

15 atmospheres. According to Swori and Ragonese (1950) the osmotic pressure in many indigenous woody plants of central Argentina fluctuates between 30–100 atm, while the Cactaceae in the same ecological condition have only 5–15 atm. This situation may be due to the fact that atmospheric moisture is the main source of water to the plant during the critical months when water stress is greatest.

The Cactaceae grow frequently in hot dry deserts, often within rocks, and their root systems penetrate to a depth of only a few decimeters. These superficial roots are fed by dew and mist, which flows along the characteristic fissures and is also absorbed by the roots, and by the stomata of the leaf-like stems.

To what extent cacti are able to absorb moisture from the air may be seen from the following experiments: Walter (1951) kept a cactus shoot for six years in dry soil in indirect sunlight, and it continued to live but lost 1/3 of its original weight. MacDouglas (1910) kept a cactus (Thewillea Somorae) dry in a museum and the plant developed new growth every summer for eight years, while its weight dropped from 7.5 to 3.5 kg during this period.

The stomatal apertures of the cacti are closed during the day and open only at night. That this opening is accompanied by more active transpiration than in daylight is proof of physiological activity at night. And, in fact, a greater synthesis of carbohydrates occurred at night.

The experiments on penetration of atmospheric moisture into the plant were performed annually during the hot summer months (July–September). All procedures connected with the growing of the plants and the transfer of atmospheric water, from the germination of the seeds to the entry of water from dew and mist into the plants, were performed out-of-doors. They were irrigated sparingly until they had reached a suitable size. The moisture content in terra rossa of pine seedlings one year old and grown in pots, was 9.2%. Groups of fifty 10–24 month old seedlings were planted in the dry sands of Rehovot during the hottest months (August and September), which are decidedly unfavorable for planting. While one group was protected with jute (the protection being removed each morning at sunrise and replaced each evening), the other was left exposed to the sun by day and to the dew and mist by night. The experiments were performed on two-year-old Aleppo pine saplings grown in pots in 1960, and in beds during 1961/63 (Gindel, 1966).

By growing plants in pairs in one receptacle (Fig. 24) we examined the absorption of atmospheric water through the leaves and its passage through the stem and roots to the soil, where it is absorbed by the roots of the partner-plant. Signs of the absorbed water were sought in the partner-plant which had not been dipped in the solutions or which had been protected against the entry of water from dew and mist. The two plants in the container were separated by a cardboard partition to prevent the direct passage of chemicals, or the dripping of water or solutions from one to the other, and so as not to contaminate the second plant with dye. Similarly, the containers were completely enclosed in plastic up to the stem level in order to prevent the absorption of water by the hygroscopic pot and also the direct passage or the dew into the soil.

The entry of water into the plant was followed: 1. by dyes—fuchsin, methyl green, methylene blue, and erythrosine, all in a concentration of 0.1%—located

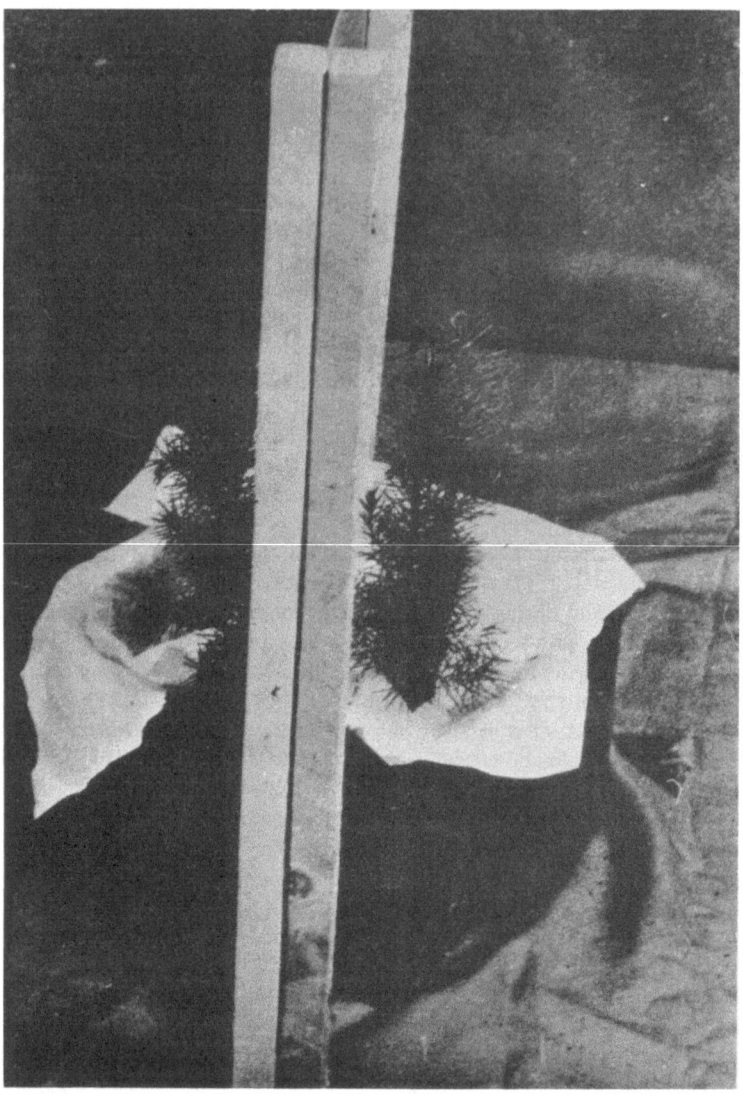

Fig. 24. Pairs of Aleppo pine seedlings in one receptacle are separated by a partition.

by anatomical examination; 2. by means of potassium permanganate which was followed by chemical analysis (oxidation); 3. by isotopes.

After half an hour's treatment of the plants in the morning with diluted dye solutions in such a way that about half the leaves were immersed (as shown in the pot reversed and the soil surface within was covered with paper to prevent the falling of particles into the bottle) the second plant was located outside the container. The plants were placed out-of-doors, after immersion so as to stimulate the movement of water. After about one hour, signs of color appeared in the undipped plants. Fuchsin left the clearest sign in maize plants, and it showed in the center of the stem and in the mesophyll and vessels of the leaves. In seedlings of Aleppo pines on the other hand, the presence of methylene blue was most marked in the phloem, and around the epithelial cells of the resin canals. Here penetration was slower than in the case of the maize.

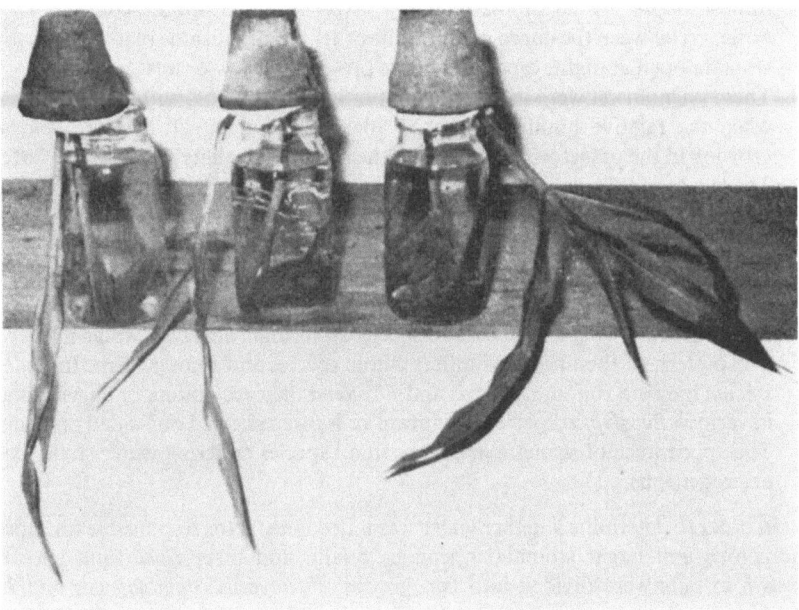

Fig. 25. Pairs of maize plants in pots, of which one is immersed in a solution and the other is not.

Here also, clear signs of dye appeared in the second, untreated plant of the pair. In this situation a high degree of soil water stress and a diffusion gradient is created. In plants that were raised in saturated soil, neither traces of dyes nor excess manganese were located in the partner plant, not immersed in the solution (Fig. 25). After obtaining positive results in the immersion experiments, the plants were exposed to the absorption of atmospheric water with their foliage powdered with dye.

Studies on the behavior of stomata during the day and night, in indigenous and acclimatized tree species, revealed the importance of absorbing atmospheric moisture in renewing a full cell turgor in the foliage, trunk and roots during the dry and hot months. (GINDEL, 1970).

Measurements of stomatal apertures of 78 tree species revealed the following information:

1. Stomatal apertures of all the 78 species which grew in dry soil with a constant water deficiency during the rainless season were larger at night than during the day. Similar results were received by HALEVY (1960).
2. Most of the stomata, or the majority of them, depending upon species and water stress within the plant, were closed during the day and particularly during the hottest hours (11:00 a.m.–2:00 p.m.). Even minimal spaces could not be observed between the guard cells. On the other hand, the same plants had many stomata open at night, especially in the presence of dew or mist.
3. These phenomena were in evidence not only during nights with dew, but even when the relative humidity reached 90–95%. Stomata of *Tamarix aphylla* growing in the desert were open when the relative humidity was only 65–70%. The leaves of this plant are covered by salt crystals, physiologically connected with the stomata, which convert water vapor into the liquid (GINDEL, 1966). (Fig. 25.1).
4. During nights with an easterly desert wind when the relative humidity dropped to 30–40%, the stomata remained closed, and there was no recovery of leaf cell turgor the following day in certain species suffering from acute water stress.
5. The pattern of the apertures differs within species and among them. In *Callistemon* they are round, in *Citrus* and *Firmania* they are oblong or curved, and in various *Eucalyptus* species the gurard cells often showed one-sided opening. The appearance of stomata apertures in all species suffering water stress was heterogeneous.

In order to determine whether water was indeed the factor responsible for more numerous and larger stomatal openings, alfalfa and three *Eucalyptus* species grown in pots were divided into two groups. *Eucalyptus camaldulensis* and *E. occidentalis* plants were grown in loess-like soil: in half the pots the soil moisture was maintained at 15.5% (field capacity) and in the second half at 7% (wilting point).* *Eucalyptus forrestiana* was grown in manured sandy soil, the soil in half the pots being irrigated when the moisture dropped below 6%. Alfalfa was grown in non-manured sandy soil, with the soil in half the pots being irrigated when the moisture dropped below 4%. The soil in the other pots of these last two species was kept saturated during the night of the experiment.

As seen in Table X, the percentage of open stomata in plots receiving the dry treatment was conspicuously greater than that in the case of both the well-watered and saturated ones. Although the two *Eucalyptus* species had a rather large percentage of open stomata when held at field capacity, those plants in the dry soil had a higher number of open stomata.

The area of stomatal opening of 10 indigenous and introduced species is presented in Table XI. In daylight the area per mm² of leaf surface ranges from

* Moisture contents of 7.0% and 15.5% were strictly maintained by the regular addition of water every few days, whenever the container weight decreased by 200–250 g.

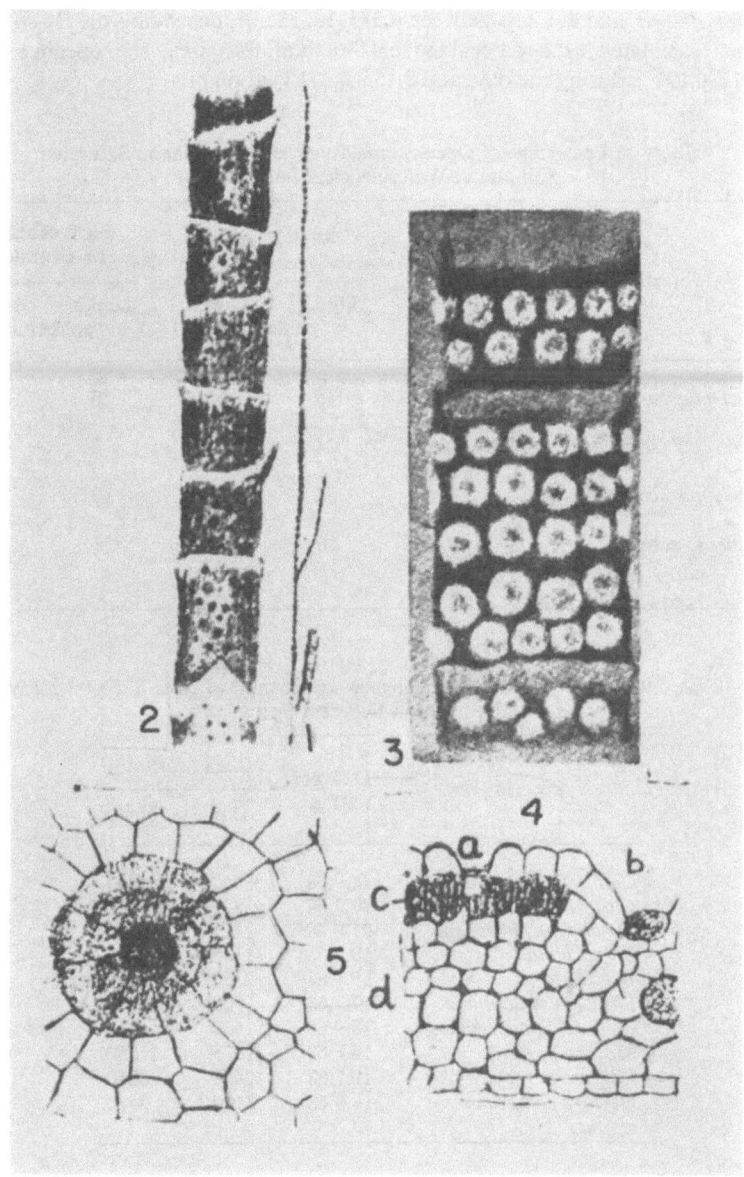

Fig. 25.1. Tamarix articulata: 1. branch; 2. section of branch; 3. sections of the branch covered with salt-crystals; 4. section through a leaf; a. stomata, b. salt-crystals c. polisade, tissue, d. water-storing cells; 5. section through epidermis in the middle is a salt gland and a crystal in the form of a hemisphere.

500–3, 450 μ^2, and at night between 4,375–13, 325 μ^2, depending on the species. When calculated as a percentage of the total leaf area, the opening were 0.017–0.122 % during the day, and 0.155–0.471 % at night.

Table X. Percentage of Open Stomata in Plants Growing in Saturated Soil and in Water-deficient Soil at Night

Species	Date of testing	Water deficit		Field capacity or saturated	
		Upper	Lower	Upper	Lower
		epidermis		epidermis	
Eucalyptus camaldulensis	2.8.68 12:00	45	50	20	25
Eucalyptus forrestiana	14.7.68 1:00 a.m.	24	29	2	4
Eucalyptus occidentalis	2.8.68 12:00	52	70	26	37
Medicago sativa	14.7.68 1:00 a.m.	16	24	0	3

Table XI. Area of Stomatal Openings in μ^2 per mm^2 of Area during the Day and Night in Different Species

Species	Date of testing	Area in μ^2/mm^2	
		Day	Night
Callistemon salignus	18.9.68	3450	4375
Citrus sinensis	17.7.68	825	9500
Eucalyptus forrestiana	21.8.68	2125	9575
E. maculata	19.8.68	1175	13325
E. robusta	19.8.68	2350	4500
Firmania platanifolia	17.7.68	700	8950
Gleditschia triacanthos	18.9.68	500	4375
Medicago sativa	14.6.68	1975	11250
Pittosporum tobira	19.7.68	3450	8250
Quercus calliprinos	14.9.68	3150	9575

According to one explanation, stomatal closure in darkness is due to the accumulation of CO_2, accompanied by a lowering of pH in the guard cells caused by the conversion of sugar into starch. This reaction leads to a decrease in the turgidity of the guard cells and to a consequent closure of the stomata. A simultaneous lowering of the osmotic pressure and a decrease of D.P.D. is a usual phenomenon. Light is the source of energy which is responsible for the photoactive movement of the guard cells. The results obtained in the present work conducted in an arid climate and in dry soil are in disagreement with this view.

STALFELT (1961), suggested that in certain circumstances, water deficit rather than light or CO_2 was the most important factor influencing stomatal movement. This has been verified by my study which produced the following information:

1. In the studied tree species suffering from water deficiency during the rainless hot season, 15–78 % of the stomata were partly open during the night.
2. By comparing the behavior of stomata in four tree species grown in saturated and in dry soil, it was shown that water deficiency was responsible for partial opening of the guard cells: there were more open at night on the lower epidermis. Our observations during nights of dew have shown that in species possessing stomata only in the lower epidermis, the opening of the guard cells at night is accompanied by a vertical orientation of the lower epidermis.

In xeromorphic plants grown in arid climates the stomata provide the only opening through which water can penetrate at night, considering the thickly cutinized and lignified epidermis which forms an absolute barrier to water movement (STOCKER, 1960). Although the apertures are small at night in the presence of dew or mist, their total area provides a significant opening for water penetration. For example, if in *Eucalyptus maculata* the sum of the openings is 0.47 % of the leaf area, then on a leaf area of 100 m² the total aperture on the lower epidermis alone would be almost 0.5 m². The problem to be solved is the rate at which dew or mist penetrates the stomatal openings, and for how long this movement continues.

The results of the present study provide an explanation of the phenomenon investigated in our previous research concerning the source of water which enables evergreen, tree xerophytes to maintain full turgor in leaf cells during the later summer months. During this period the roots exhaust the available soil water, and an absolute or a partial *status quo* of soil moisture exists from June or July until the rains begin in October or November.

It is still premature to lay down a rule on the amount of water attracted by the xerophytes from the air during different months of the year. This quantity is influenced by several factors, apart from species, such as the deficit in the soil, the extent of physiological activity, soil type, fluctuations in dew and mist, etc.

However, the following equation gives a basis for calculating the amount of atmospheric water absorbed per hectare of forest:

$$S = (R + W) - (E + T)$$

where

W = the amount of atmospheric water attracted by the plant

E = physical evaporation within the forest due to the operation of the thermodynamic factors which operates here with a lesser intensity than in bare areas. Physical evaporation included also the interception of rainwater by the forest canopy.

T = amount of water lost by transpiration

R = yearly rainfall

S = excess of soil moisture in comparison to the bare area.

The appreciable quantity of air-borne dust during the prolonged dry season was found in other arid places in the world (HUTTON, 1958) as well; this has led us to believe that such atmospheric water must also bring with it different ions, eliciting the opinion that it consists essentially of distilled water. However, analysis of dew and mist performed during the dry months confirmed the above mentioned viewpoint. Table XII presents the results of four dew and two mist analyses

Table XII. Chemical Composition of Dew and Mist (mg/1) collected during October–November, 1963 (Rehovot)

Ions	Mist		Dew			
	25.0	27.10	20.10	30.10	21.11	30.11
Chloride	70.9	38.3	191.9	21.3	24.1	8.51
Sulphate	162.1	75.7	96.3	37.0	61.7	39.5
Nitrate	1.0	1.3	0.4	1.9	4.9	1.4
Bicarbonate	48.8	67.0		54.9	62.2	81.9
Calcium	38.1	50.1	54.1	32.0	40.0	35.7
Magnesium	8.5	4.9	3.6	4.0	4.9	1.7
Potassium	15.3	10.4	9.0	7.0	6.8	3.8
Sodium	27.6	15.9	5.6	5.8	11.4	9.6
pH	6.5	6.9	6.35	6.8	6.9	6.9
Total dissolved solids	400	245	—	155	210	150

collected in Rehovot during October–November 1963. It shows the presence of anions and cations of nutritional importance; potassium and large amounts of the sulphates of calcium, magnesium and sodium relative to their chlorides. Traces of fluorides, nitrates, ammonia and metasilicic acid were also found; the six samples were acidic.

The chemical constitution of dew and mist, as detailed in the preceding paragraph, raised four questions:

1. Are the ions, found in the atmospheric water, absorbed by xerophytes through the stomata?
2. Do these ions enter the soil as a result of the exudation of part of the atmospheric water by the roots?
3. What is the relationship between fluctuations in water content and ionic concentration as a function of depth?
4. What is the difference in amount and distribution of the ions in forest soil and identical exposed area?

To clarify these questions, an experiment was performed in a 14-year-old tamarisk forest in dunes near Revivim. On August 8, 1963 samples of soil were taken from 20 places in the forest and outside it, to a depth of 240 cm, according to the method previously described for examining soil humidity. Figure 25.2 shows

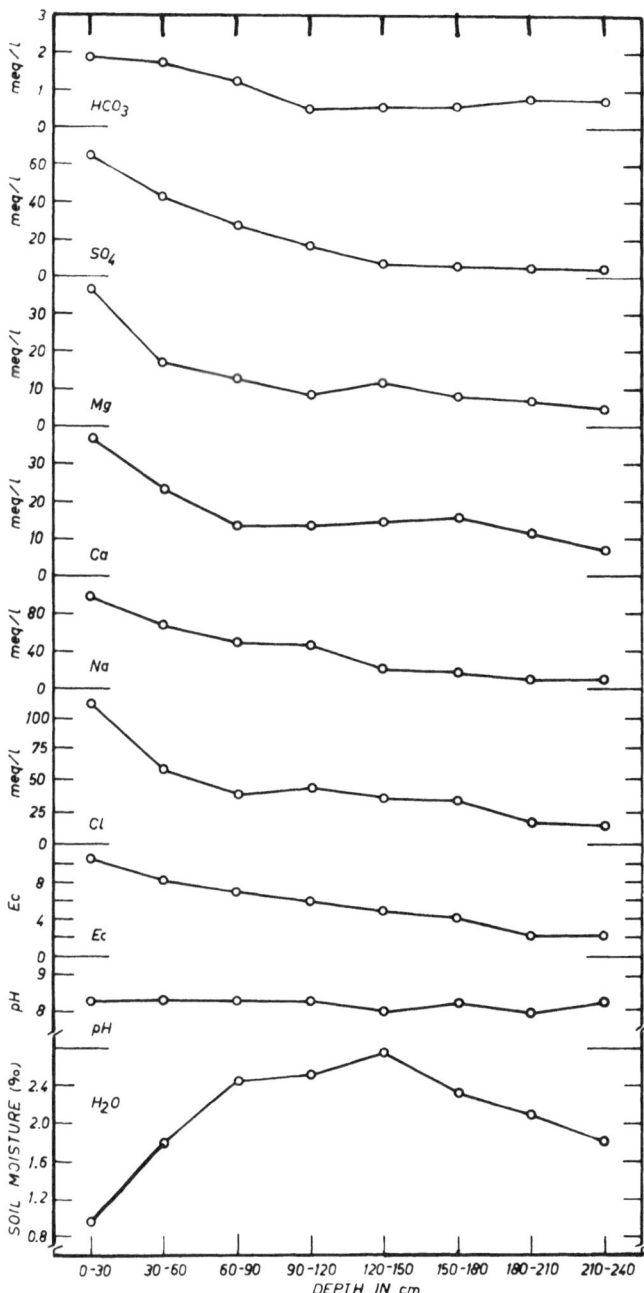

Fig. 25.2. Tamarix aphylla forest. Fluctuations in the ionic concentration and soil moisture percentage with depth. Revivim (desert).

the results of electrical conductivity measurements of 6 ions and the pH at which different layers were examined; further, it shows the fluctuations in percent soil humidity with depth. The figure shows that the ionic concentration decreases with depth and reaches a minimum at 210–240 cm, with a maximum concentration in the 0–30 cm layer. The moisture percentage on the other hand, is minimal in the upper soil layer and reaches a maximum actually in the 120–150 cm layer, where the ionic concentration is approaching a minimum; however, the soil humidity is still relatively high (1.8%) at a depth of 240 cm.

Tamarix, a typical halophyte, when growing on non-saline sandy soils or dunes, collects salt from the air. If the composition of the ions within the soil under the Tamarix canopy are compared to an open non-forested area, the changes caused by halophytes in sand will be clear. Table XIII shows that there is an increase in the four elements observed. Some light ion increase was noted also in the two glycophyte sepecies *Pinus halepensis* and *Eucalyptus camaldulensis*.

Table XIII. Fluctuation in the Ionic Concentration
a. under the forest canopy; b. and in open area (Gevulot)

Species	Na		HCo₃		Ca		Mg	
	a.	b.	a.	b.	a.	b.	a.	b.
Euc. camaldulensis	4.79	3.70	3.57	2.21	7.8	8.7	3.93	3.58
Pinus halepensis	6.02	3.12	2.22	2.07	9.40	9.69	2.60	2.31
Tamarix aphylla	9.85	3.42	5.07	2.42	11.40	5.71	5.20	2.66

The soil moisture regimen and the role of atmospheric moisture in woody xerophytes presented in the former paragraph awakened the following problems analyzed in the next paragraphs: 1. What is the pattern of the consumption of the two water resources, rainwater and atmospheric moisture, which the woody xerophytes exist from? 2. What is the pattern of the transpiration process, particularly during the critical months when a *status quo* in the quantity of soil moisture has been reached? 3. To what degree is transpiration correlated with and even parallel to other physiological processes? and finally, 4. What is the ecophysiological definition of the transpiration process?

IRRIGATION OF WOODY XEROPHYTES WITH ATMOSPHERIC WATER WITHIN THE DESERT

The desert soils are suitable, so far as the climate and the quality of the soil are concerned, for the growth of many forest tree species which can occupy most of the area but the expansion of which is prevented solely by the lack of water. This shortage has given us the idea of exploiting water from dew and mist of irrigation of forest plants, and of concentrating the little rainwater available into the immediate area of the roots.

There are many species of xerophytes which can be utilized for afforestation even in desert regions where only 100–200 mm of rain fall annually, if they are irrigated for the first 2 or 3 years after planting. This conclusion is based on the results of afforestation in the Negev Desert over the past 20 years, and now woods, avenues and windbreaks of tamarisk, eucalyptus and Aleppo pine lend a touch of green to the gray, arid desert vista.

By exploitation of water from dew and mist, which is actually greatest in the desert, and by concentration of the rainwater in the root region, as is described in this paper, it is possible today to plant endemic and domesticated xerophytes woody species at every place where such sources of water are available (GINDEL 1965).

The investigation recorded here was limited to the subtropical zone of Israel and to those desert regions possessing the following characteristics: 1. A total annual rainfall of 25–200 mm (Beersheba–Eilath), limited to the cooler season. 2. The rainfall is limited to 8–10 days in dry years and 20–40 days in normal years. During the rest of the year, the sky is clear and the number of hours of sunlight is close to a world maximum. 3. Extreme annual fluctuations in the rainfall and frequency of drought years. 4. Scarcity of streams and rivers, apart from a gravitational flow from the mountains into the depressions and plains. This flow ceases after a few days as the water reaches the Dead Sea and the Red Sea. 5. An almost complete absence of sweet water suitable for drinking or for the use of plants.

Although there are local variations, there is a general distribution in the number of dewy nights (ASHBEL, 1949) as follows: 1. An increase in the number of nights with dew and its quantity going southwards in the country; an annual variation between 5 mm—Dead Sea region and 150 mm—Northwestern Desert. 2. There are, in fact, the greater number of dewy nights during the hottest and driest months. 3. In the Western Desert, where the rainfall is lowest, the number of nights with dew reaches 250 mm annually (Gvuloth), 140 of these with "heavy" dew. 4. Along the Mediterranean coast, in the Central Hula Valley and in the west-central part of the Valley of Israel the dew is heavy. 5. There are about 160 nights with dew in the hills with fluctuations depending on height and exposure; there is least on the eastern slopes.

GILEAD and ROSENAN (1954) measured the number of nights with dew and the quantity collected at 80 meteorological stations in the country at the same time, over a period of 10 years, by means of DUVDEVANI's apparatus (1947), and reached the following conclusions: 1. The number of nights with dew is more or less constant, and the fluctuations are similar to those found in the rainfall. 2. There was no constancy between the various stations, and the fluctuations were not only influenced by thermodynamic factors, but also by soil structure and topography (hills). 3 More dew is attracted by plants grown in light soil than in clay soil. 4. That measured in plantations was about half that in nearby exposed areas.

Amount of "Flowing" Dew and Mist per Unit Area

Some nights were completely dewless; on others, although drops condensed on the sheet, the amount was insufficient to lead to a flow through the gutters into the bottles. The flow of water from dew and mist into the measuring bottles of the apparatus begins when the sheet has become appreciably covered with drops. Table XIV indicates the "flowing" water collected per square meter of polyethylene. This flow differed in 3 places belonging to different climatic regions

Table XIV. Monthly Amount of "Flowing" Dew and Mist (cc) per Square Metre of Polyethylene in 1964

Month	Location		
	Rehovot	Revivim	Eshtaol
July	1,148	1,608	1,222
August	1,648	2,653	2,048
September	876	1,643	1,555
October	2,056	3,631	2,684
November	862	1,270	988

Table XV. "Flowing" Dew and Mist (per cent) from the Northern and Southern Slopes of the Instrument in 1963

Month	Rehovot		Location Eshtaol		Revivim	
	North	South	North	South	North	South
August	50.41	40.59	55.8	54.2	60.7	39.3
September	54.72	45.28	47.9	52.1	64.5	35.5
October	51.58	48.42	47.9	52.1	59.9	40.1
November	61.18	38.82	49.1	50.9	61.5	38.5
December	60.64	39.36	—	—	50.52	49.48

(Revivim—desert, Eshtaol—subtropical hills, and Rehovot—semi-arid plain), with most at Revivim and least at Rehovot.

The apparatus was arranged with its two planes facing north and south, respectively, and here, also, differences emerged. Table XV shows the greater flow from the north face at Revivim and Rehovot, and from south face at Eshtaol. In the summer of 1964, however, the southern aspect gave a greater flow at Rehovot.

Irrigation of Forest Saplings with Atmospheric Water

In 1962, 8–10-month-old saplings of the following species were planted at Rehovot and in the Judean Hills (Hulda): *Pinus halepensis, P. brutia, Cupressus horizontalis, Eucalyptus gomphocephala. Tamarix aphylla* seedlings were planted in dunes south of Beersheba. Near each plant was arranged a plastic sheet, 1.3 m² in area (Figs. 15, 16). The dew was led to the plant-pit through a plastic gutter.

The experiments were repeated in the summer of 1963 and, apart from those saplings planted in the rainy season, others, of the same species, were also planted in August, when the conditions (heat and drought) are unsatisfactory for survival. One group was provided with atmospheric water by means of the plastic sheets, while the other was planted nearby without such aid. While the former survived until the winter rains, the latter had dried up by the beginning of September

Concentration of Rainwater by Means of Plastic Sheets

The same plastic sheets which had proved useful for the collection of dew and mist were also used for the collection of rainwater. Table XVI shows the slight and unreliable rainfall in the desert over the past 18 years with fluctuations from 32 to 177 mm at Revivim, and from 100 to 255 mm of Gvuloth. These figures also include a number of light showers of 0.5–1 mm, which only wet the upper 1–2 cm of the soil. This moisture evaporates within a few days, owing to the strong sunshine between the showers even during the rainy season. By concentrating this rainfall, its effectiveness can be increased. For example, a sheet of 5 m² area supplying a pit of 1 m² area may provide 5 times as much water as when water falls on a similar pit which is not provided with a sheet and, consequently, a 5-fold percolation. In fact, the percolation depends on the physical properties of the soil, especially its permeability, which is greatest in dunes, and least in loess.

The collection of atmospheric water and the trapping of rainwater, by means of hygroscopic material (polyethylene or polyvinyl chloride), increase the sylvicultural potentialities of desert and semi-arid regions.

It was found that 20–30 liters of atmospheric water flowing into the plant-pit from each square meter of the polyethylene roof was sufficient for the continued existence and for some growth of the tree xerophytes used in this work, during the first summer after planting. It is possible, also, to plant woody xerophytes, such as Aleppo pine, Tamarix sp., and Eucalyptus sp. and others, even in midsummer (July–August) when the amount of dew reaches maximum, and they can be maintained without wilting so long as they are provided with moisture from dew and mist until the advent of the rains. In such places as the deserts of Chile and Peru, for example, this is even more important, since the mists may reach an annual value of 300–400 mm, while the actual rainfall does not exceed 40 mm.

The arbori-sylviculture possibilities in the desert may also be increased by the concentration of the little rainwater available in the region of the root. Here, also,

Table XVI. Yearly Rainfall (mm) and Number of Rainy Days at Two Locations

Year	Location	
	Revivim	Gevulot
1953/54	113.2(27)	—
1954/55	92.0(19)	148.8(19)
1955/56	130.3(36)	177.0(39)
1956/57	177.5(34)	224.5(42)
1957/58	32.4(22)	100.3(25)
1958/59	120.5(32)	142.3(22)
1959/60	44.1(16)	—
1960/61	104.4(32)	111.5 (8)
1961/62	71.8(26)	162.1(24)
1962/63	32.8(8)	48.9(20)
1963/64	164.9(33)	248.7(38)
1964/65	188.0(37)	308.0(35)
1965/66	86.0(28)	129.0(26)
1966/67	123.0(37)	197.0(35)
1967/68	130.0(39)	221.0(41)
1968/69	88.0(26)	138.0(26)
1969/70	55.0(17)	107.0(25)
1970/71	136.0(27)	227.0(31)

the efficiency depends on the area of the sheet, and it can be adapted to the species. If, for example, a certain region would allow the growth of any forest species so far as the soil and climate are concerned, but the annual rainfall is only 200 mm instead of the 1,000 mm necessary, it is possible to increase the rainwater available by leading the run-off from a 5 m² sheet into a plant-pit of 1 m² area.

ACCEPTED TRANSPIRATION CONCEPTS

DARWIN found a linear relationship between transpiration and relative humidity. CURTIS (1926) defined transpiration as a "necessary evil" and opposed the view that any benefit can come from it. He believed that transpiration could only bring harm to the plant. He thought that transpiration rates could exceed the rates of evaporation from open sources of water.

Unquestionably, moisture is necessary for plant growth, but according to MILLER (1938) and others, transpiration, from a theoretical viewpoint, may or may not be a necessary condition for plant survival.

Most of the water absorbed from the soil by the plant is lost by transpiration without participating in physiological activities. Moreover, the same weather conditions that control the rate of evaporation also affect the rate of transpiration.

WEAVER (1949) regarded evaporation by way of the stomata as essential for the living plant which must remove excess water from the protoplasm and cell-walls.

BONNER (1955) compared the plant to a wick, drawing water from the soil and leading it via the stem to the stomata of the leaves, through which it evaporates. KRAMER postulated approximately the same ideas on transpiration. Evaporation of water through stomata is often so pronounced that leaves wilt at mid-day. Transpiration hinders growth and causes the death of thousands of saplings in the forest.

Clothes on a line dry rapidly when the relative humidity is low, the temperature high, and the day windy. Likewise, transpiration in plants is rapid under such conditions (NORTHERN, 1953). SLATYER (1959) defined transpiration as a passive process determined by 1. atmospheric conditions, 2. the amount of water in the soil relative to the atmosphere, and 3. the energy gradient along the transpirational pathway created as a result of evaporation from the leaf. He believed that transpiration can continue even after the death of a plant, as it is more a mechanical than a biological activity. So also ECKARDT (1960) viewed the forest as a water reservoir, and the foliage as a surface from which moisture evaporates in response to the intensity of the climatic factors acting upon it.

According to GATES (1966), the process of transpiration is necessary to cool the leaf and to defend it from overheating. "If the plant physiologist", says GATES, "ever has had a doubt concerning the value to the plant of transpirational cooling as it affects leaf temperature, this gross misconception should be dispelled now once and for all."

KRAMER (1969) distinguishes two principal sites of evaporation from a leaf, one located in the walls of the mesophyll cells bordering intercellular spaces, the other in the outer surfaces of the epidermal cells. He claimed that transpiration had a harmful effect and was an unavoidable evil, and because of it much water is consumed until turgor is lost in the leaf cells and many branches wilt. Furthermore, the function of transpiration is to facilitate water movement and to prevent the leaf from overheating.

It has been a popular view, therefore, that transpiration is a plant weakness which cannot be avoided, owing to the unfortunate structure of the leaf. Transpiration is virtually a process of evaporation (SALISBURY and ROSS, 1969), modified somewhat because of the structure of the plant, and, although the stomata occupy but 1 % of the area of the leaf, diffusion of water vapor through them may often reach 50 % or more of the rate of evaporation over open water sources. A fall in transpiration at noon occurs even in well-watered plants; owing to the inability of the plants to absorb water as fast as it is lost.

This summarizes the trend of opinions from the beginning of the 18th century until today.

Some physiologists, meteorologists, and physicists have considered as transpiration only the water vapor diffused from the leaves. However, quite substantial amounts of vapor are diffused from other organs of the tree having stomata and lenticels, for example, leaf pedicels, branches, twigs, the stem, the flowers, the fruits (HUBER, 1956). In *Phragmnites communis* transpiration through the stem sometimes approaches 3–6 mgs/g/min (KINDLE, 1953). The degree of participation of the green stem in gaseous exchange and photosynthesis increases in some xerophytes under severe climatic conditions, and reaches a peak in the desert.

Excretion of water through the lenticels of the bark is also considerable, even though their area may not exceed 2 % of many tree species. In some it amounts to 20 % of the amount transpired through the leaves (HUBER, 1956). Also, after reaching a particular phase in their development, fruits often participate in the exchange of gases, and even help to balance a shortage of water. This was shown in fruit-bearing apple trees which survived a drought better than neighboring trees which had been stripped of their fruit (KRAMER, 1949).

The quantities of water given off by plants at night may be as much as 5 %, or even 10 %, of daylight transpiration (DAUBENMIRE, 1959).

Winter transpiration is a non-negligible amount (5–15 %) of water exhaled from trees. Its relative quantity varies under arid conditions depending on temperature and physiological activity. In *Pinus taeda*, for example, winter transpiration can amount to about 14 % of summer transpiration (KRAMER and KOZLOWSKI, 1960).

Since 1932, with the appearance of THORNTHWAITE's meteorological approach to transpiration which identifies it with physical evaporation, the biological meaning of that process has been progressively ignored. THORNTHWAITE and HARE (1955), concluded that the rate of transpiration is a function of 1. the external supply of energy to the evaporating leaf surface, principally by solar radiation, and 2. the capacity of the air to remove vapor, i.e., wind speed, turbulent structure, and the decrease of vapor concentration with height.

In spite of a few dissenting voices, it is still contended that the plant community is a reservoir of water which evaporates according to the intensity of the climatic factors, particularly solar energy, which according to the physio-meteorological concept is the main force acting on the evaporation surface. And, according to SLATYER (1968), in the presence of sufficient water, total evapotranspiration from a plant community, expressed per unit land area, may exceed that from a similar area of bare wet soil or free water.

VAN DEN HONERT (1948) and PENMAN and SCHOFIELD (1951) viewed leaves as passive evaporating surfaces because the calculated resistances of open stomata, they thought, were small compared with external air resistance. VAN BAVEL'S (1968) opinion that the crop canopies are passive evaporative surfaces better known as wicks was disputed by LEE (1968). To LEMON (1966), transpiration is a simpler process controlled largely by purely physical factors.

However, it follows from our studies on transpiration of woody xerophytes as a function of environmental factors and other physiological processes that transpiration has another implication.

TRANSPIRATION DURING THE SEASON OF GROWTH

Heterogeneity in the constellation of the climatic factors in the arid areas, and in the available soil moisture both seasonally and annually, leads to conspicuous differences in the intensity of transpiration. Figures 26 and 27 show mean monthly transpiration of 3 planted Aleppo pine forests, one on Mt. Carmel and two in the Judean Hills. The rate of transpiration* gradually decreased beginning in April, and reached its minimum value of 100–200 mg/g/h, during the period from August

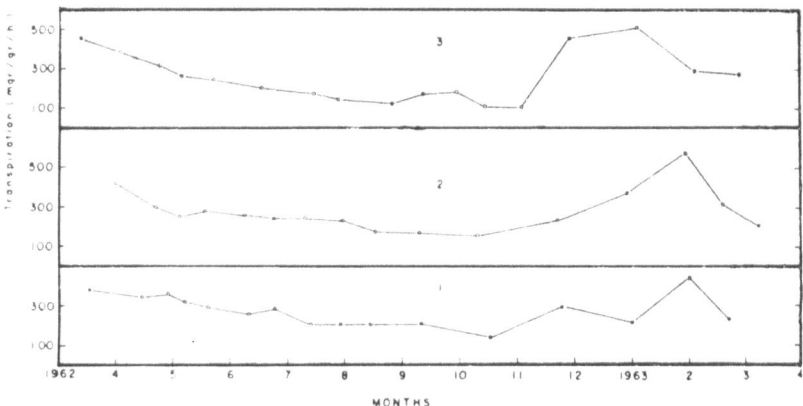

Fig. 26. The progress in transpiration rate (the sum of six trees) of the Aleppo pine in three planted forests: 1. Eshta'ol. 2. Hulda. 3. Carmel.

to November. The high transpiration rate during January–February, 1963, is correlated with unusually high temperatures for these months, and with soil moisture availability.

Figure 28 shows the variations in the course of transpiration in three xerophytes (*Eucalyptus camaldulensis, Pinus halepensis*, and *Tamarix aphylla*) grown in the desert in sandy soil (GEVULOT). The year 1964/65 was one of the most rainy years of the decade, while the year 1965/66 was a typical drought year. This accounts for the differences displayed in transpiration during the critical summer months

---

* Calculated for the group of 6 dominant trees measured each hour in each experimental plot (GINDEL, 1969).

of 1965 compared to those of 1966. The drought affected Eucalyptus more severely than Pinus. The Tamarix was scarcely affected for reasons which are discussed below.

From Fig. 28 it is evident that transpiration rate is lowest during the months December–January or February, depending on species, although at this time of

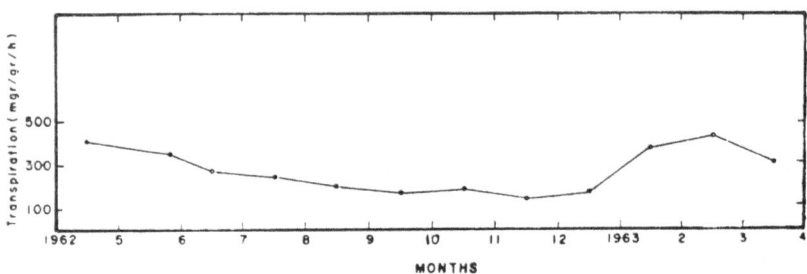

Fig. 27. Average transpiration of the three planted forests: 1. Eshta'ol. 2. Hulda. 3. Carmel.

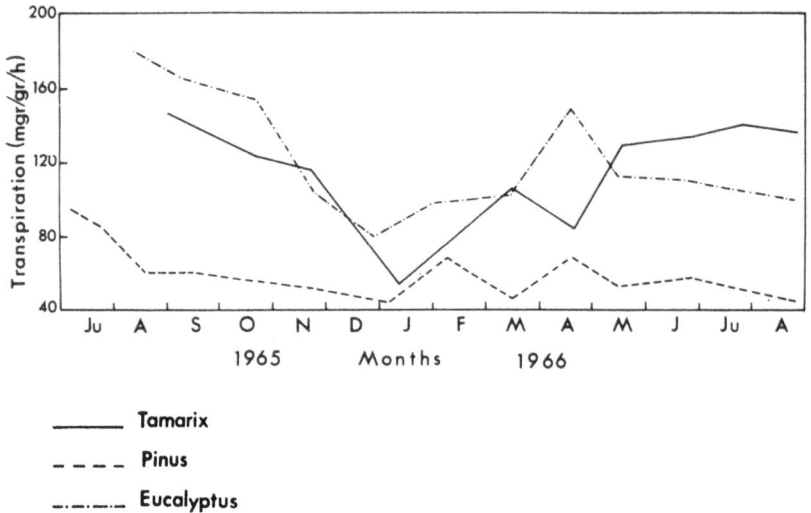

Fig. 28. The monthly fluctuations in transpiration rate in three tree xerophytes during 1965/66. Gevulot.

year the soil moisture content is higher than in summer. Tamarix was found to be most vulnerable to lower temperatures, even during the period of physiological activity. In April, 1966, for example, when the average day temperature was 20.3°C the transpiration rate decreased to 83.3 mg/g/h, while in March and May, when

70

temperatures were above 28°C, transpiration rates were 105.1 and 129.5 mg/g/h, respectively.

The transpiration rate of Pinus in the hills of Judea facing the desert was higher throughout the summer than at the Gevulot grove. The average difference was, however, only 19.5%, while during the year in question the amount of rainfall recorded in Lahav was almost twice as great as in Gevulot. The average transpiration rates for all species for the period of the experiment revealed that the highest average rates were found in Eucalyptus and Tamarix (122.5 and 122.4 mg/g/h, respectively) as compared with 61.3 mg/g/h for Pinus.

In some instances transpiration varied with crown aspect. In Tamarix, transpiration from foliage on the south-facing side was 2.5% higher than on the north

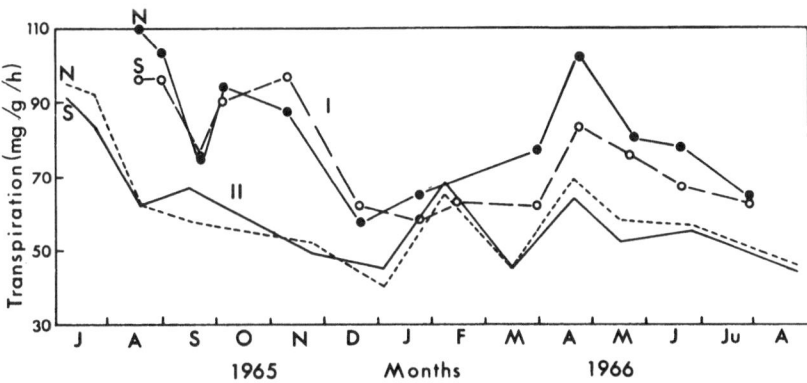

Fig. 29. Variations in transpiration rate (per tree) measured on northern (N) and southern (S) sides of the foliage in Lahav (Southern part of Judean Hills) I and Gevulot II.

side, and in Pinus it was about 1% greater on the south aspect. In Lahav, transpiration rates on the north-facing side of Pinus were 7% higher than on the south side (Fig. 29). South and north sides gave similar values in Eucalyptus. The experimental period included the summer of 1965, which was preceded by a rainy year (308.5 mm at Gevulot), and the summer of 1966, which was preceded by a drought year (129.4 mm). This explains the differences in transpiration rates measured during those two summers in Pinus and Eucalyptus.

The march of transpiration in the sub-tropical and desert zones shows that transpiration is a minimum when it is needed most for cooling of leaves: During the hottest and driest months, transpiration drops to its lowest values. This contradicts the prognosis of GATES (1963).

The pattern of transpiration for Aleppo pine during the day is characterized by large fluctuations from hour to hour. Figure 30 depicts transpiration from each of the three young forests. Two characteristic days were chosen for the measurement in each forest: one at the beginning of the summer when soil moisture and transpiration were high, and the other at the end of the summer when the soil was dry.

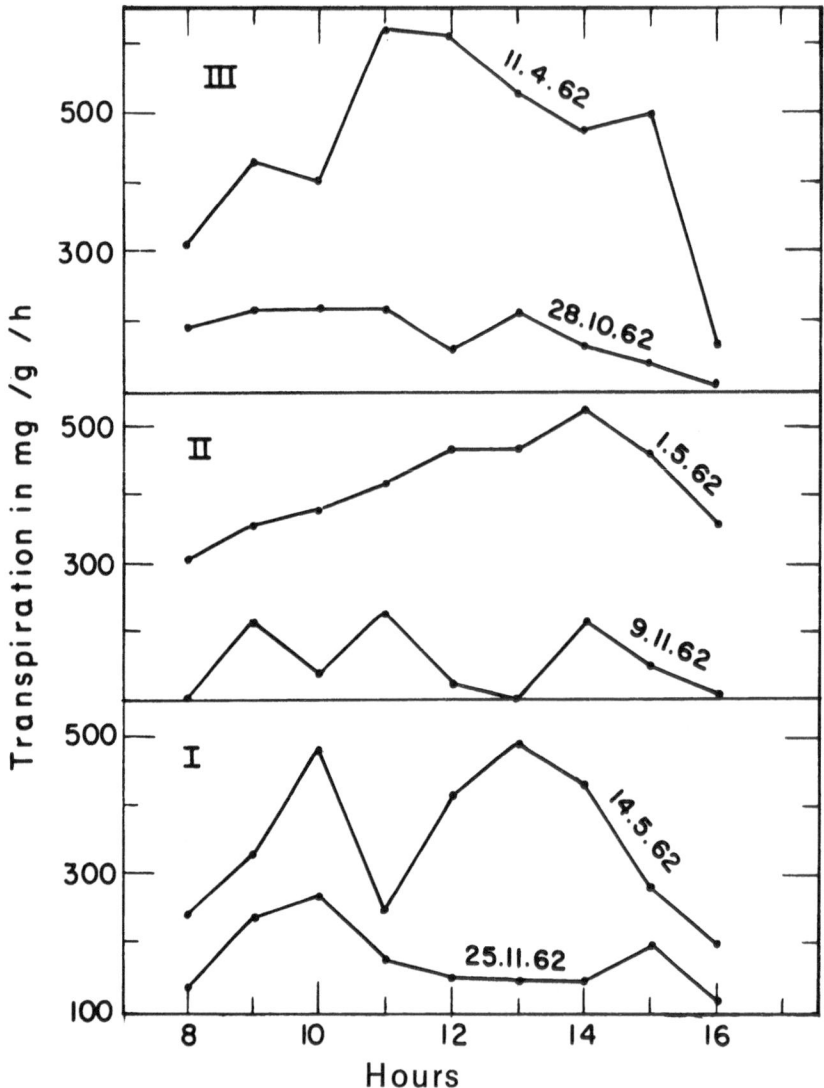

Fig. 30. Characteristic fluctuations in hourly transpiration rate (the sum of six trees) in three forests: 1. Eshta'ol. 2. Hulda. 3. Carmel.

The curves are different in the two cases. The lowest transpiration values occur at 0800 and 1600 hours in all of the forests. There are a certain number of days during the year, however, in which the absolute minimum occurs in Esthaol at 1300–1400, in Hulda at 1300–1500, and on Carmel at 1000–1200, and very seldom at 1300–1500 hours. Differences in rates of transpiration at the beginning and end of the dry season are conspicuous.

The investigation has shown that, although some trees (or a certain side of the tree), on some days, may show a drop in transpiration at a specific hour, such mid-day depressions are not always observed. On another day the rate may rise gradually to a maximum and then decline toward evening without a drop at noon.

TRANSPIRATION AS A FUNCTION OF THE FOLLOWING ECOPHYSIOLOGICAL FACTORS

A. *Physical Evaporation, Temperature, Relative Humidity, and Wind Velocity*

In mesic or humid climates, in the presence of sufficient water in the soil, various workers have obtained a parallel between the rates of transpiration and evaporation, finding that heat and wind evaporate water from foliage at the same rates as from open water. According to them, wind decreases the concentration of water vapor at the leaf surfaces, and thus increases the rates of transpiration. STALFELT (1956), found that the leaves of *Betula* often show a rate of transpiration equivalent

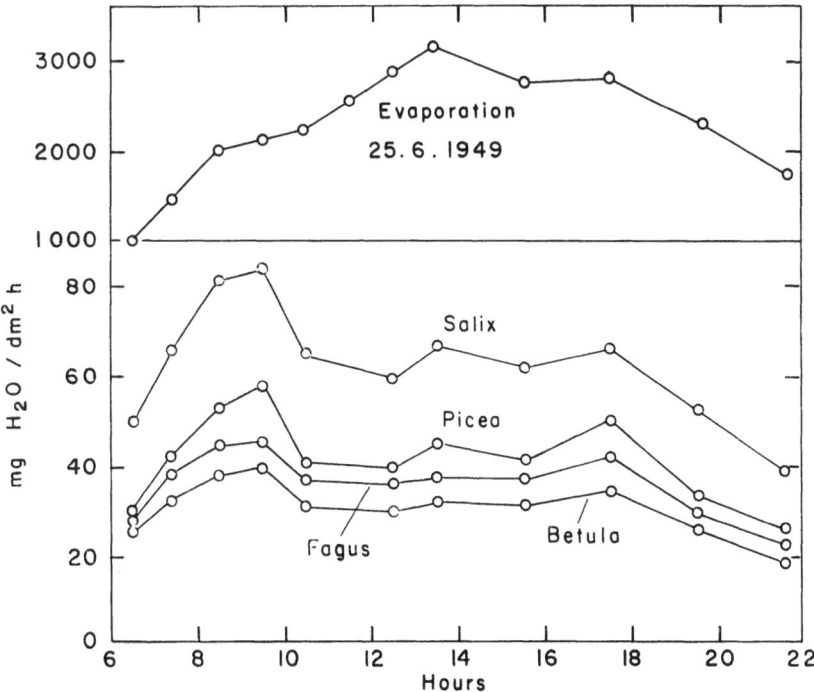

Fig. 31. Daily fluctuations in evaporation and in transpiration from the bark of 4 species of trees growing in a temperate climate, according to HUBER (1956).

to only 65% of evaporation from open water and according to his calculations the diffusion of water through the stomata, per unit area of open stomata, sometimes is 50 times as great as open water evaporation.

The correlation of physical evaporation and transpiration has led to the use of a combined terminology. "Evapotranspiration", for example, is the sum of transpiration from plants and evaporation from bare soil. THORNTHWAITE and HARE

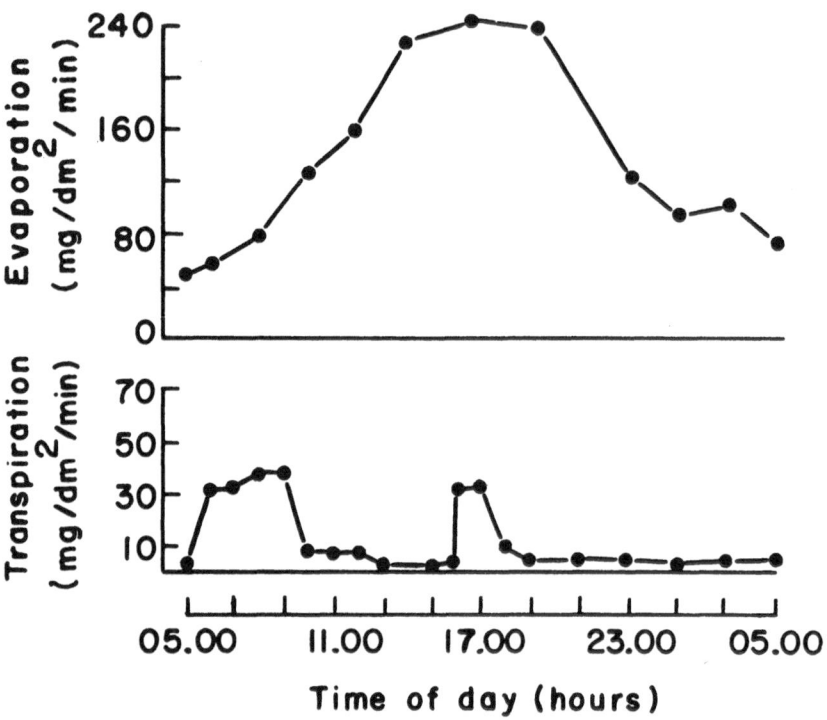

Fig. 32. The hourly fluctuations of evaporation and transpiration in *Rhagodia baccata* of young, medium-sized leaves during 1965/66 (Lakeside, Western Australia) HELLMUTH, 1968.

(1955), explained transpiration on the same thermodynamic basis as physical evaporation, and calculated its rate, using DALTON's law and similar formulas. They calculated potential evapotranspiration based on the mean temperature and latitude. PENMAN (1948) used other variables, i.e., radiation, air temperature, humidity, and wind velocity. These are the four factors that are known to govern the rate of physical evaporation. He claimed that the plant can be regarded as a passive channel between the water in the soil and the atmosphere above.

The similarity of responses to atmospheric stress between transpiration and evaporation disappears as climatic and edaphic conditions become more severe.

In subtropical conditions in Israel, for example, the rate of transpiration decreases even in irrigated orange groves during hamsin days in spite of the fact that the rate of evaporation is then three times as great as on other days (HALEVY, 1956). But even in mesic conditions, the linear relationship between the rates of transpiration and evaporation holds during the growing season only when reserves of water are available in the soil. When the water reserves in the soil decrease, transpiration decreases. Figure 31 illustrates this for four tree species growing in a cold temperate climate. While maximum evaporation occurs at 2:00 p.m., maximum

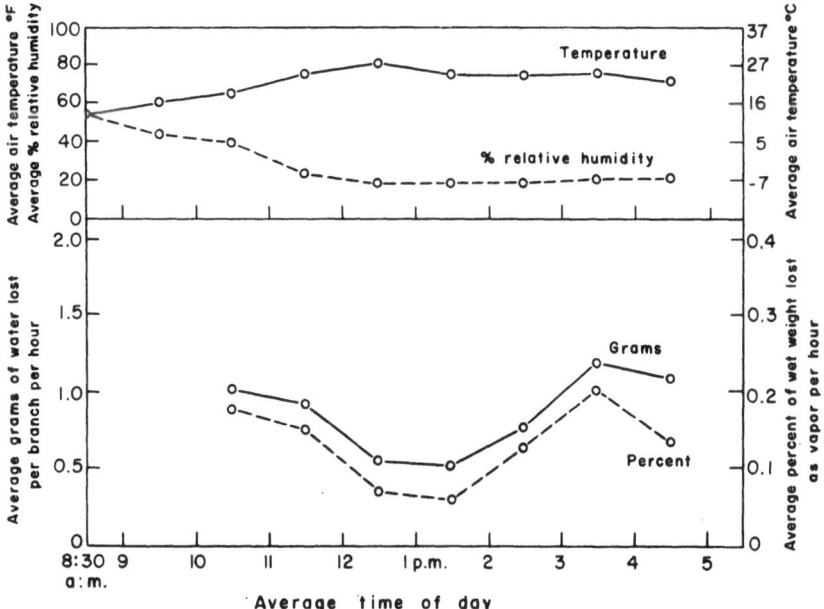

Fig. 33. *Sarcobatus vermicultatus* (Hook) Torr. The hourly fluctuations in water loss from branches on a sunny day (9th September 1965) in relation to temperature and air humidity (STARK, N. 1967).

transpiration occurs at 9:00–10:00 a.m. according to the species. The parallel between transpiration and evaporation disappears even in rainy climates as the shortage of water in the plant rises. In fact, in many experiments during the hours of maximum evaporation, transpiration, is at a minimum.

As shown by the studies of POLSTER (1950), also in the temperate climate of central Europe there is no parallelism between fluctuations in transpiration and physical evaporation in the three predominant tree species: *Betula, Quercus* and *Picea*. This is understandable under conditions in which the temperature and the winds encourage physical evaporation but prevent physiological activities from occurring at the same rate.

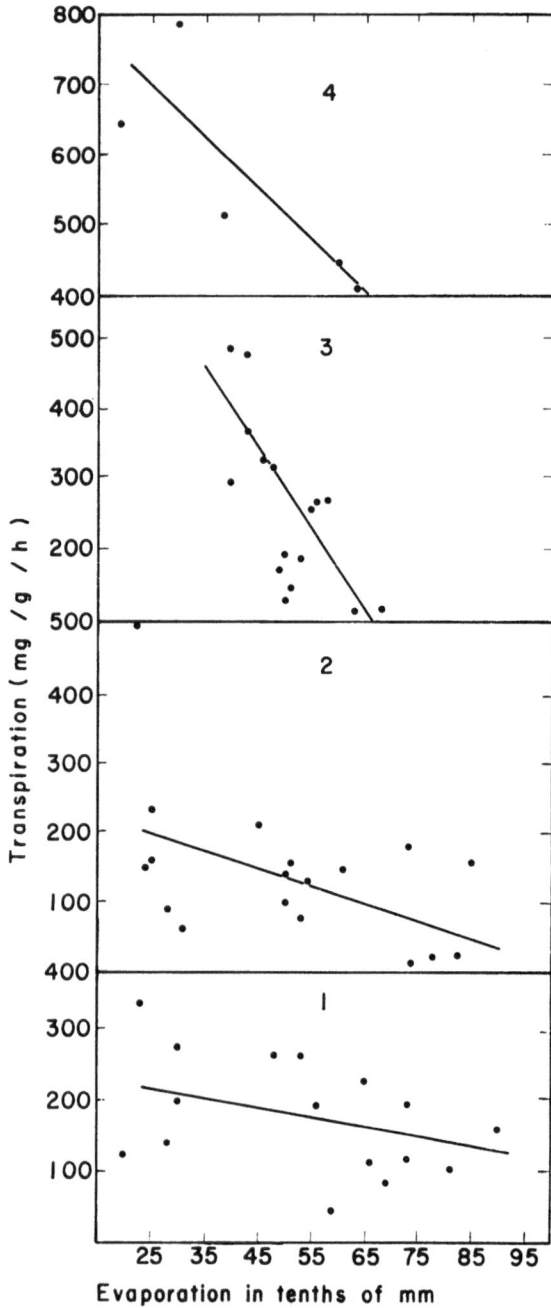

Fig. 34. Transpiration rate (the sum of six trees) as a function of evaporation intensity: in Aleppo pine forests: 1. Hulda. 2. Eshta'ol. 3. Carmel. 4. Cana'an.

GRIEVE and HELLMUTH (1970) found the same relationship between physical evaporation and transpiration during the dry and hot season in some indigenous woody plants in Western Australia (Fig. 32). This biological phenomenon also appears in other arid regions as shown by investigations carried out by STARK (1967) in the Death Valley, California. The water loss from branches of Sarcobatus vermiculatus was greatly reduced during the hottest hours of a sunny day in September, in spite of the fact that relative air humidity dropped to its lowest values (Fig. 33).

Correspondence between the physiological activity, transpiration, and the physical process of evaporation, is limited to certain hours of the day and particular seasons of the year. Even in moist climates, the relationship becomes obscured as a shortage of water in the soil limits the activity of the plant, in spite of more favorable conditions of light, temperature, or wind which increase rate of evaporation

There is a negative relationship between the rate of transpiration and the intensity of evaporation in xerophytes (GINDEL, 1967), when standard multiple regression techniques are used to calculate the mathematical relationship between them (Figs 34, 35).

$$y = a_1 + a_2 x$$

y = mean hourly transpiration

x = evaporation in tenths of mm.

The negative correlation results from the fact that high evaporating conditions are associated with hot dry weather or strong winds, whereas both of these lead to a simultaneous diminution of transpiration intensity.

A similar negative relationship was found between transpiration and intensity of temperatures (Figs. 33, 35).

Figure 36 shows the connection between temperature and transpiration in *Fagus* and *Larix* (according to EIDMAN) growing in a hilly region in a temperate climate where the winter temperature often falls to $-20°C$. The data are for June when the trees reached full physiological activity. When the temperature fell below 15°C, the rate of transpiration reached a minimum; above 25°C, it attains a maximum. The optimum temperature for increased transpiration in mesic plants generally varies between 20–30°C. For *Betula*, optimum transpiration occurs at 20°C, and for *Pinus* at 30–32°C.

There is also a difference in the temperature range that facilitates the activity of the roots, as opposed to that of the parts of the plant above soil-level. According to KRAMER (1942), optimal transpiration and water absorption occurred in soil at 25°C for 4 species of pine. As the thermophilic character of the plant is more pronounced, the optimum temperature for heightened transpiration rises. The more northern the habitat of the species, the lower is the optimal soil temperature for transpiration; it reaches a peak in hot deserts. Transpiration increases with soil temperature even at 40°C if sufficient water is available.

When the soil temperature drops below a certain minimum value (which varies with the species), the roots cannot absorb water owing to a rise in its viscosity, and in the viscosity of the protoplasm. The viscosity of water is twice as great at 0°C as at 25°C.

During the hottest months of June–August in cold or temperate climates when the plant is physiologically most active, there is liable to be a correlation between the different physiological processes and evaporation since in these months temperature reaches maximum values. As pointed out in the previous chapters, the relatively moderate temperature in these climates leads to intensive physiological

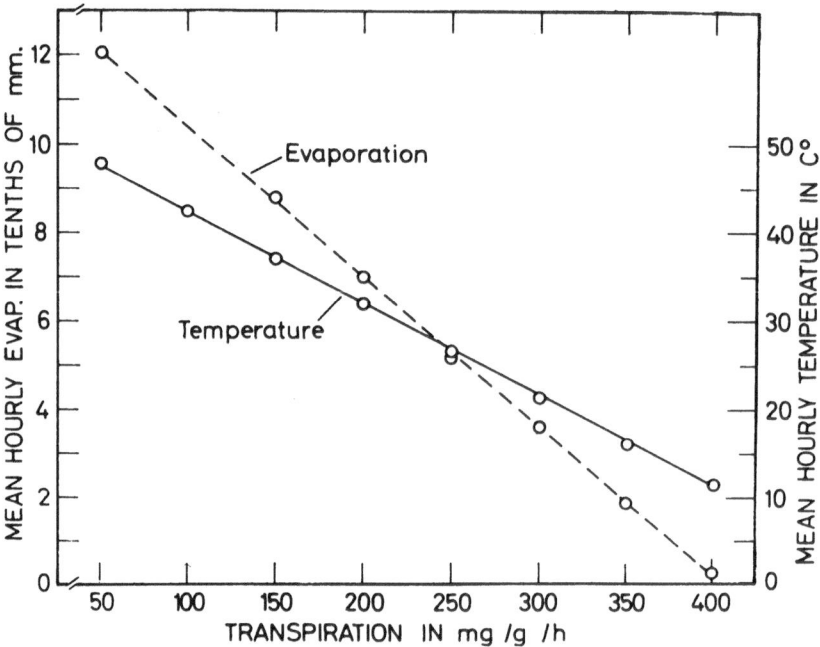

Fig. 35. The progress of transpiration (the sum of six trees), evaporation and temperature, during the years 1962–1963 measured in six Aleppo pine forests grown in the hills of Judea, Efraim and Carmel.

activity, accompanied by a high transpiration rate. The extent to which temperature is a decisive factor in regulating plant activity in a cold climate is indicated by the results obtained by SWANSON (1966) who studied the water flow in trunks, a process comparable to the transpiration rate. This investigator compared two tree species (*Picea engelmanii* and *Pinus contorta*) endemic to the Rocky Mountains at an elevation of 10,000 ft where the annual rainfall is about 600 mm, with about 1/3 falling in the summer months. Figure 37 shows that the water movement in the trunk reaches a maximum rate in the months May–September accompanied by a high rate of growth. Also the work by GINDEL (1954), who examined one of the predominant species (*Abies alba*) growing in the Apennine Mountains, showed

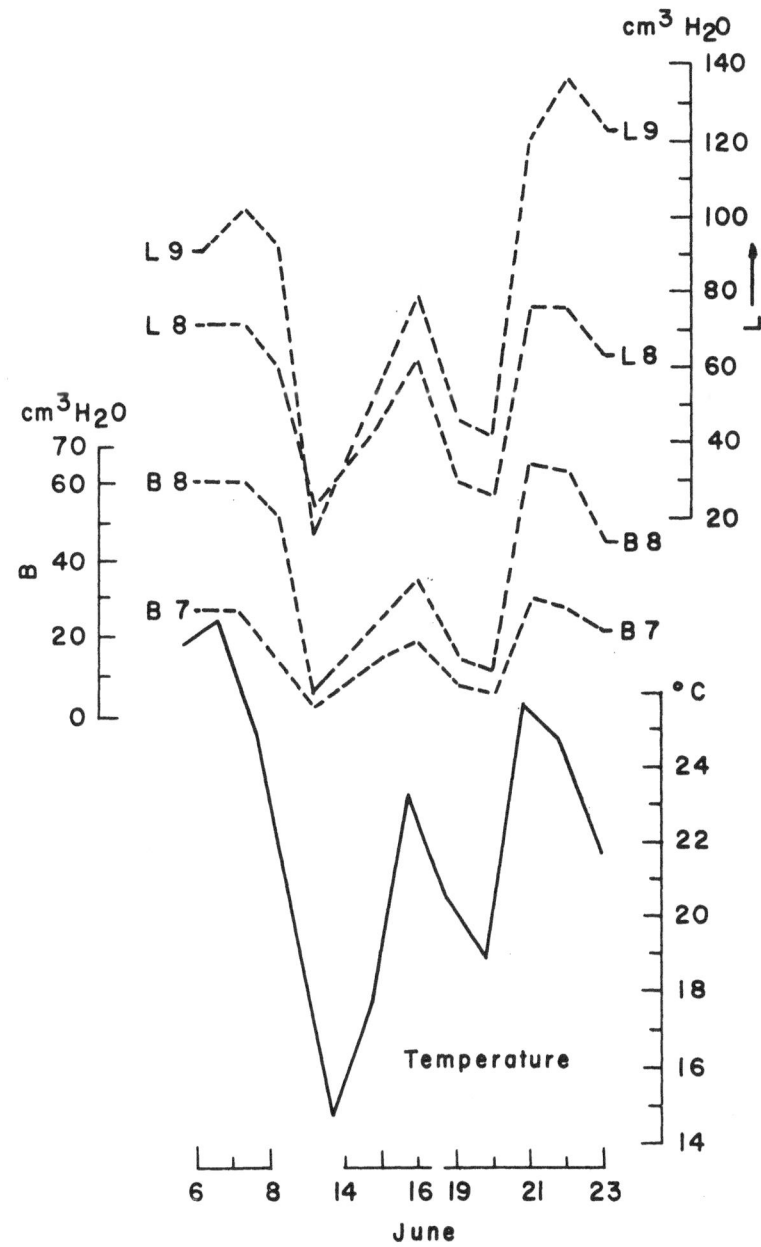

Fig. 36. Daily fluctuations in mean temperature and the rate of transpiration in 2 Larix trees (L8 and L9) and 2 Fagus trees (B7 and B8) (According to Eidman).

81

that in addition to the temperature which stimulates rapid growth in the months June-August, the rainfall of less than 150–200 mm produces only slight growth, whereas the rains in the fall and winter months are not significantly correlated with annual growth.

It follows that the temperature most suitable for transpiration varies in thermophilic, megathermic, and microthermic plant species. Plants growing in the cool wet season continue to absorb soil moisture at a lower temperature than do those in hot climates. In elephant grass, the power of absorption of the roots declines at 10°C, and in cotton at 17°C.

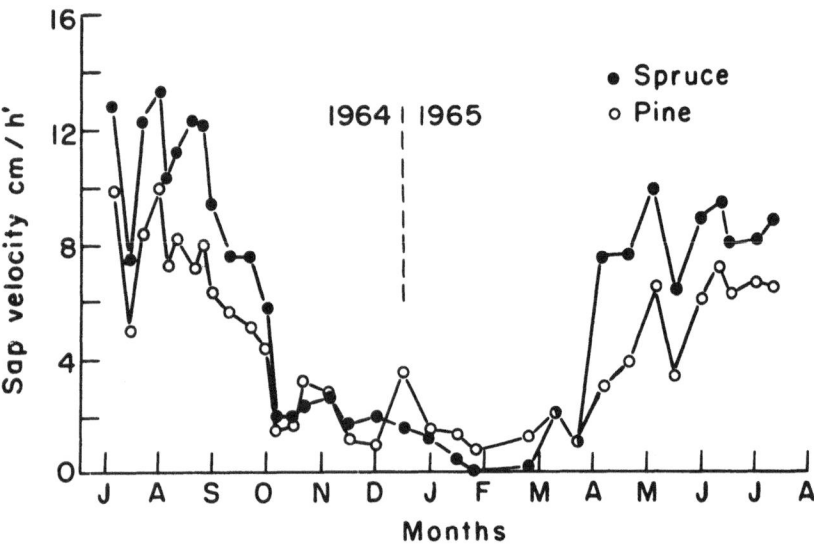

Fig. 37. Manually-read sap velocity in Engelmann spruce and Lodgepole pine at mid-day on clear to partly cloudy days. Each point is the average during the time of reading, usually one hour or less (SWANSON, 1965).

It is not temperature alone which raises the rate of transpiration, but rather temperature has its optimal value if moisture is available. The biological properties, such as the degree of thermophily, and the favorable action of other ecological factors, are the controlling influence.

During the physiological resting phase, although the wind may be maximum, and evaporation high, transpiration is at a minimum, and is too small to be measured. For example, coniferous trees in winter in a mesic climate have a rate of transpiration equal to 1/55–1/250 of that in the summer, and 1/300 in the following species: *Picea excelsa, Picea obovata, Pinus sylvestris,* and *P. cembra.* These low rates occur in spite of abundant soil moisture.

Wind velocity during the months November–March in different parts of the

United States fluctuates between about 5 and 60 miles per hour. These values are in many cases higher than the wind velocity during the summer months when plant activity is high. During the period of strong wind velocity in winter, when plant activity is nil or very weak, evaporation may vary from 1–6 inches monthly, while transpiration drops to a minimum. This is another clear evidence that transpiration is not governed by the same rules as evaporation. Nevertheless, strong winds during

TRANSPIRATION AND SIX ECOLOGICAL FACTORS

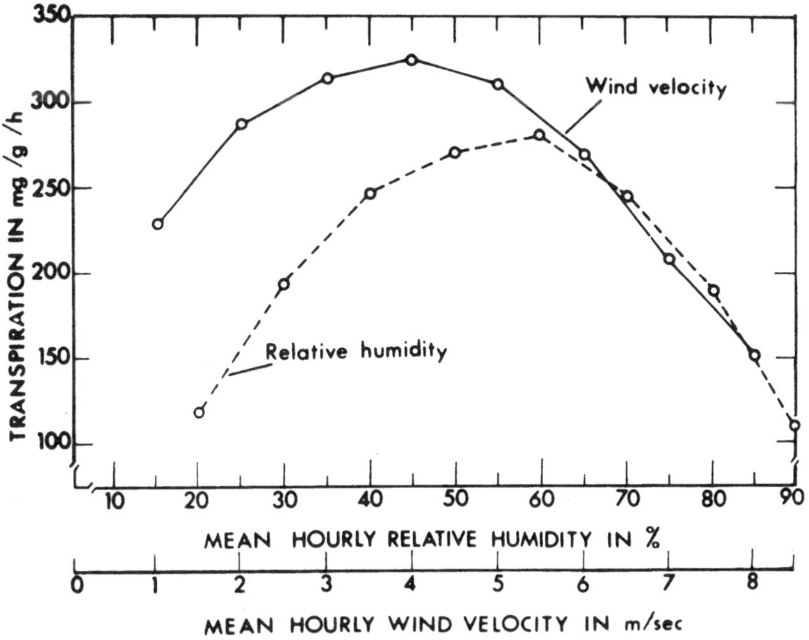

Fig. 38. The march of transpiration (the sum of six trees), wind velocity and relative humidity Aleppo pine.

the period of growth, which may influence the whole metabolic framework, affect transpiration too.

When a water shortage in the soil intensifies and physiological activities cease, the rate of transpiration falls in any case, no matter what the strength of the wind. Under conditions of extreme water shortage, increased winds may further weaken transpiration and even hasten its cessation. NAKAJAMA and KADOTA (1948), STOCKER (1956), examined the rate of transpiration during a wind storm and found that it was equal at both sides of the plant, although the wind velocity was 6 meters/second to windward and only 0.13 meters/second to leeward; the rate of transpiration at both sides was 2.6 mg/dm²/min.

STOUTJESDIJK in his physiological studies on more than 30 tropical and temperate plant species found that generally a very strong influence of wind on transpiration is rarely to be expected as the bigger part of total leaf resistance is situated inside the leaf and is unaffected by wind speed.

The effect of wind velocity and relative humidity on transpiration in Aleppo pine forests is expressed by a parabola (Fig. 38):

$$y = a_1 + a_2x + a_3x^2$$

x = mean hourly relative humidity in % or wind velocity in m/sec.

When the wind velocity is less than 5 m/sec there is a positive relationship; above this figure transpiration declines.

A similar connection exists between transpiration and relative humidity, with a positive relationship until 50–60% humidity (Fig. 38).

B. *Transpiration as a Function of Soil Moisture Depletion*

Monthly fluctuations in soil moisture and transpiration in the hills of Judea and Efraim are presented in Figs. 39 and 40. In Eshtaol soil moisture decreased until the beginning of August and transpiration until July.

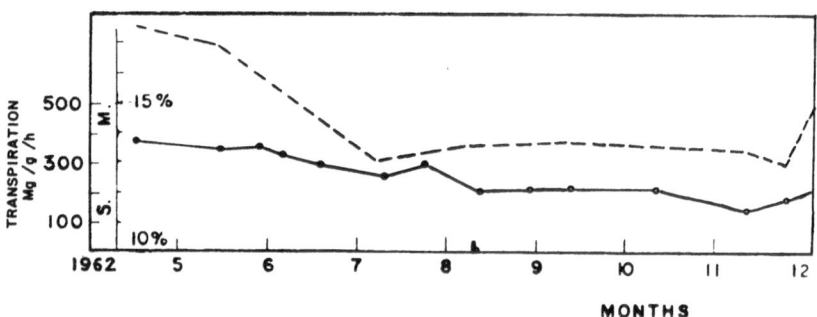

Fig. 39. Transpiration (the sum of six dominant trees) (solid line), soil moisture (S.M.) (broken line) (Eshta'ol). 1962.

In the Hills of Efraim (Fig. 40) transpiration and soil moisture drop rapidly, in parallel fashion, from March–April until the end of June. Transpiration continued at a rate of 131–191 mg/g/hr until the rains, in spite of the fact that the soil moisture from the end of June until the 19th of September diminished by one-half percent only. The curve of soil moisture which is the mean for three 30 cm layers fell during the period of the end of March until the end of June from about 18% to 12.6%. The mean transpiration rate for the same period is 261 mg/g/hr. During the following 3 months (July–September) the rate was 150 mg/g/hr, and the edaphic moisture should have dropped by 2.7%; but in fact it dropped only 0.5%. The remaining 2.2% of moisture required to provide the above transpiration rate must have thus come from another source. Since groundwater is a long way from the roots, and since the soil water percentage in the neighboring bare area was less than under the canopy, the only possible source of this excess water is the dew and mist adsorbed by the foliage (GINDEL, 1966).

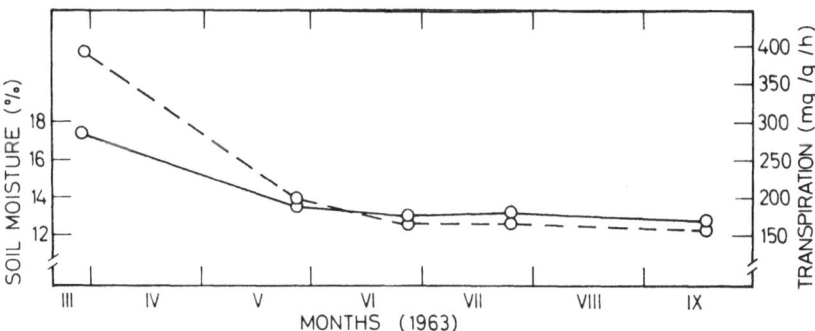

Fig. 40. The fluctuations in transpiration rate (the sum of six trees) and soil moisture percentage in the Hills of Efraim. Aleppo pine.

C. *Transpiration as a Function of Solar Energy, Evaporation, Soil Moisture and Leaf Area*

Eucalyptus camaldulensis Dehn., *E. gomphocephala* A.D.S. and *E. occidentalis* *Endl.* successfully acclimatized in the subtropical and northern desert zones of Israel have been used for this study. The experiment was carried out in Rehovot which belongs to the sub-tropical region.

Field experiments were carried out in a natural environment with only soil moisture under control.

Water was applied to the soil surface in the pot, in an amount calculated to bring the soil moisture content to 15.5%, which is equivalent to the water retention at 1/3 atmosphere.

In March 1968, all the foliage was cut and 30 cm of leafless main stem was left in each pot, the aim being to obtain new foliage ontogenetically homogeneous. Then one hundred seedlings of each species were divided into two groups, one kept at the same moisture percentage as before cutting (15.5%) and the second in a water deficiency of 7%.

While the seedlings of *Eucalyptus camaldulensis* and *E. occidentalis* produced a similar foliage when kept in the same soil moisture percentage, the mean leaf area of the newly developed foliage grown in water deficiency changed after cutting.

It diminished to between 31–43% depending upon species and season of testing (Table XVII).

The influence of increased soil moisture on two groups of eight months old *E. gomphocephala* seedlings grown in sandy soil is presented in Table XVII. One group was kept in a steady water deficiency and the other at field capacity. Transpiration was measured on October 20, 1966 and October 21, 1966. In the first group of the mean, leaf area was reduced to 60 and 68% and the transpiration rate to 23.2 and 38.9% respectively on the two days. The transpiration of the water deficit plants increased to almost the same level as that of the plants kept at field capacity when soil moisture was raised to an identical percentage. The differences for leaf area and transpiration rate were significant on 1% level.

Similar results were obtained in the two other Eucalytpus species. (*E. camaldulensis* and *E. occidentalis*.) On September 2, while the rainless hot season still continued, the plants grown under a water deficiency had a 43% smaller leaf area and a lower transpiration rate. They showed however an increase by 36% in transpiration on September 6 after irrigation up to field capacity. The mean hourly evaporation was 0.45 mm and 0.38 mm. and the mean hourly global solar radiation 58 and 50 cal/cm^2, respectively, on the two days (Table XVII).

On November 14, 1968 and January 16, 1969, the same phenomenon was repeated, i.e., an increase in soil moisture for plants grown in water deficiency changed the transpiration rate to the same level as (and slightly above) that of those grown at field capacity (0.01 significance level). The low transpiration during January is due to the more thermophilous character of *Eucalyptus camaldulensis* in comparison to *E. occidentalis*. The global radiation during these two days was half that in the two former days (2.9 and 6.9). In *E. occidentalis* a similar situation followed as a result of replenishing the soil moisture of the plants grown under water deficiency. In this species, the mean hourly transpiration of the plants grown at field capacity was by about 9% higher on December 31, 1968 than on the hot dry summer day September 3, 1968.

From Figure 41, which depicts the hourly fluctuation in transpiration, it is evident that in *E. camaldulensis* the maximal rate approached 400 mg/mm^2/h during 2.9 and 6.9, while during November 14, 1968 and January 16, 1968 it dropped to about 320 and 240 mg/mm^2/h, respectively. By contrast, *E. occidentalis* showed a maximal rate of 850 mg/mm^2/h during the rainy cool season (December 31, 1968), after soil moisture was replenished. In this instance too, transpiration was higher (by 34%) than in the group grown at field capacity (0.01 level). The hourly fluctuations in global solar energy do not parallel those of transpiration.

From Table XVII and Fig. 41, it is evident that in spite of the smaller mean leaf areas of all 3 species grown in water-deficient soil, their transpiration intensities equaled or even exceeded those of the plants grown at field capacity, when soil moisture in the former was raised to the same level. This occurred even though both groups had been exposed to the same intensity of solar radiation. A study of the number and length of stomata in both groups revealed that those grown in a water deficient soil are from 16 to 21% more numerous.

It follows that a xeromorphic leaf characterized by a smaller area and greater density of stomata can increase its rate of transpiration two-fold or more,

Table XVII.

Mean Leaf Area (cm², A), transpiration rate (mg mm^{-2} h^{-1}, B), Global Radiation (cal cm²) and Evaporation (tenths of mm) in three Eucalyptus species

Species	Date of testing	Steady water deficiency		After irrigation to field capacity		Steady field capacity		Mean hourly global radiation	Mean hourly physical evaporation
		Leaf area	Transpiration	Leaf area	Transpiration	Leaf area	Transpiration		
E. gomphocephala	Oct. 20–66	8.6	133	8.8	560	14.3	569	—	—
	Oct. 21–66	8.2	214	9.5	549	12.0	575	—	—
E. camaldulensis	Sept. 2–68	11.5	100	—	—	20.1	237	58.2	0.45
	Sept. 6–68	—	—	8.0	366	14.5	232	50.1	0.38
E. camaldulensis	Nov. 14–68	9.4	102	—	—	13.6	231	34.4	0.29
	Jan. 16–69	—	—	9.1	130	12.5	108	29.6	0.16
E. occidentalis	Sept. 3–68	6.0	256	—	—	9.1	370	60.2	0.42
	Dec. 31–68	—	—	10.5	541	15.6	407	31.1	0.28

independent of the intensity of solar radiation. This will occur as soon as soil moisture is replenished to the same level as in the well-watered plants. It occurs because solar energy, similar to other ecological factors, is a link in the assembly of the climatic and endaphic factors. Only optimal values of radiation (and not its

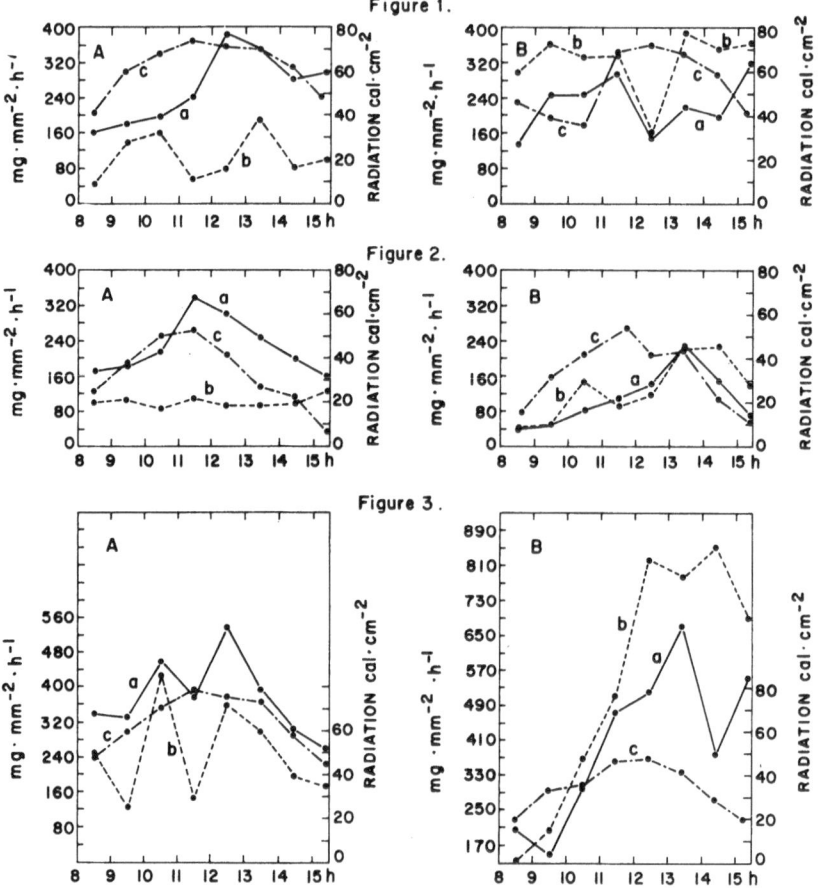

Fig. 41. 1. *Eucalyptus gomphocephala.* 2. *E. camaldulensis.* 3. *E. occidentalis.* Transpiration (mg/cm²/h). Curves (a) at field capacity, (b) in water deficiency, (c) solar radiation (in cal/cm²).

intensity), which depend on the biological properties of the species, led to a maximum transpiration rate. Indeed, a 36% higher transpiration rate in *Eucalyptus occidentalis* was obtained during the cool season, when solar energy was only half and mean hourly evaporation less by 33.4% of that measured during the summer day, 6.9.

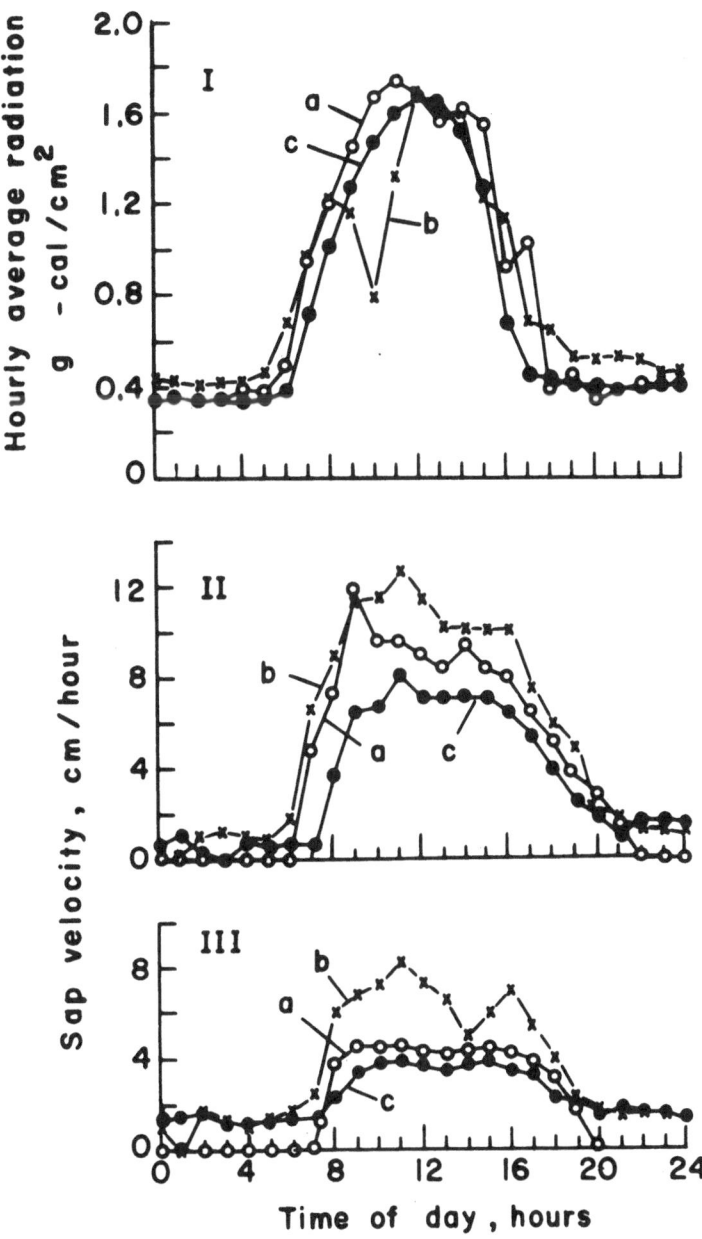

Fig. 42. Fluctuations in hourly radiation (I), sap velocity in Engelmann spruce (II) and Lodgepole pine (III) during 15.6.65 (a), 22.7.64 (b) and 3.9.64 (c). SWANSON, (1965).

The lack of a parallelism between transpiration rate and radiation intensity is also evident from the studies by SWANSON (1965), who examined this problem on two prevalent tree species in their Rocky Mountain habitat (Engelmann spruce and Lodgepole pine), at an elevation of 10,000 ft above sea level. In this cold climate, the radiation intensity was almost similar during June, July and September (Fig. 42), ranging from 17.0 to 17.5 $g.cal^{-1}. cm^{-2}$. On the other hand, the rate of water flow in spruce trunks dropped from 12.4 cm h^{-1} in July to 8 cm h^{-1} in September. The decrease in pine trees was even more drastic: from 8 cm h^{-1} in July to less than 4 cm h^{-1} in September. This condition was due to the drop in temperature in September, since there was an adequate supply of soil moisture from the summer rains.

D. *Transpiration as a Function of Leaf Moisture*

Fluctuations in the moisture percentage of leaves and the rate of transpiration was studied in the following species: *Pinus brutia, P. canariensis, P. halepensis, P. pinea, Eucalyptus camaldulensis, E. gomphocephala, E. occidentalis,* and *Tamarix aphylla.*

In Fig. 43, monthly variations in transpiration rate and leaf moisture in two pine species grown in the neighborhood of Eshtaol are compared. Transpiration in Aleppo pine is highest in April, whereas in *Pinus brutia* it is maximal at the beginning of March. This may be ascribed to the fact that the Aleppo pine is the more thermophilic and higher temperatures are required for this phenomenon. Leaf moisture in this species also is maximal in March. The mean transpiration rate is 34% less in *Pinus brutia* than in Aleppo pine for the total measurement period. In Aleppo pine a much lower hydration in the foliage is accompanied by a higher transpiration rate, whereas in *Pinus brutia* the reverse is observed. This fact is a well-known ecophysiological occurrence. It is expressed by the so-called "transpiration-ratio" which is the ratio of water used to dry matter produced.

Differences in percentage leaf moisture for the two pine species over the observation period were significant at the 0.01 level, with a S.E. of 0.546. When transpiration rates were compared, differences were significant at the 5% level, S.E. = 0.203.

A comparison of the same two parameters for the species *Pinus halepensis* grown at two different sites is presented in Fig. 44. One grove was planted in the desert among sand dunes near the Gevulot settlement, while the second was grown in the Mediterranean maqui at the southern boundary of the Judean Hills (Lahav), facing the desert. Despite the very shallow layer of soil above the rock at Lahav, the fact that the annual rainfall is at least twice that at Gevulot, and that the climate is more moderate due to the northern exposure of the slope, faster growth accompanied by a higher rate of transpiration occurred at Lahav.

In Fig. 45, data are presented for two neighboring groves, each composed of a single species: *Tamarix* indigenous to the desert, and *Eucalyptus* acclimatized from the subtropical region of Australia. Both species were grown in sandy soil in the desert (Gevulot). In this case an opposite situation existed: The differences in leaf

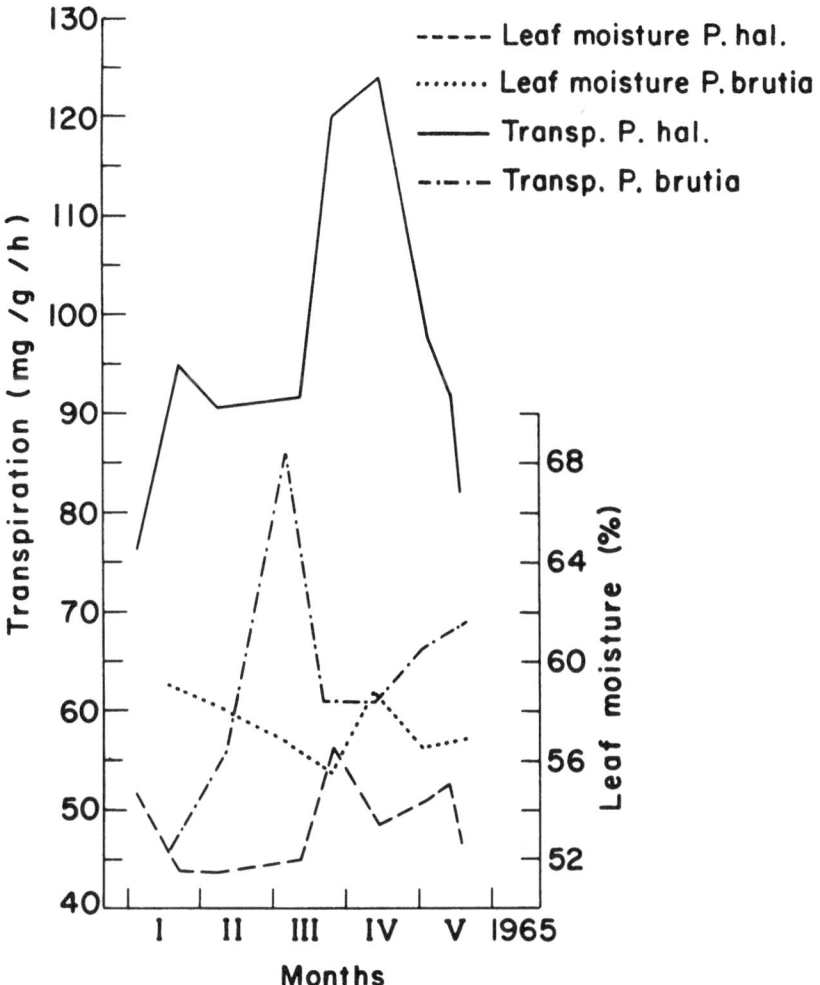

Fig. 43. Percentage leaf moisture and transpiration rate in two neighboring groves of *P. halepensis* and *P. brutia* (Eshtaol).

moisture between the 2 species were conspicuous, but were small in regard to rate of transpiration. The mean percentage leaf moisture in *Tamarix* was 62.78 % and in *Eucalyptus*, 51.15 %; the difference was significant at the 1 % level, S.E. = 1.089. Maen transpiration rate differences are non-significant, S.E. = 8.80.

Nor were conspicuous differences in the fluctuations of water content percentage in the leaves of woody xerophytes noted in the Death Valley by STARK, (1969), as follows: Atriplex 66–84 %, Larrea 50–80 %, Pancephyllum 68–97 %.

From the data presented in Table XVIII, further conclusions can be drawn: a. the mean transpiration rate as well as percentage of leaf moisture show only small differences in all the three locations studied. b. Percentage mean leaf moisture was quite high—55.0%. c. Differences between maximum and minimum values of leaf moisture percentages in all the Aleppo pine groves ranged from 7.3%–20.0%. d. The greatest difference between the highest and lowest rate of transpiration is found on Mt. Carmel where transpiration during the summer months is 17.66 mg/g/h, a five-fold diminution from the maximum, while leaf moisture decreases only by 9%.

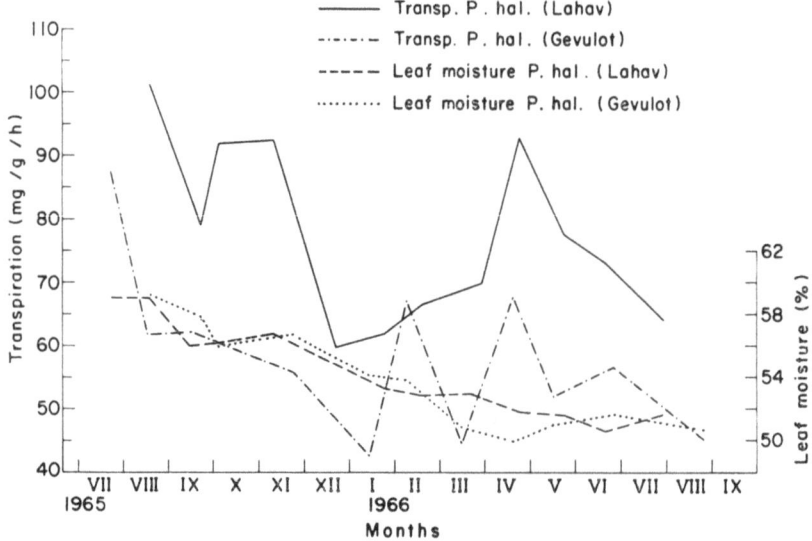

Fig. 44. Percentage leaf moisture and transpiration rate in Aleppo pine growing at Gevulot and at Lahav.

The reaction of the tree xerophyte to lack of water in arid regions is *not* evidenced as a decrease of water percentage in the leaves to a level leading to a conspicuous state of dehydration. Rather, there is a decline in physiological activity, including transpiration.

Due to the reduction of water tension in the plant and the return of turgor in the cells by the absorption of atmospheric water by the foliage during the hottest and driest summer months, transpiration continues to a small extent in spite of the *status quo* in soil moisture. So, in spite of the fair hydration of the leaves during the critical months, transpiration declines to its lowest values.

The opinion that water can move at low rates for long distances to plant root by unsaturated flow without detectable changes in soil water content (GARDNER,

92

Table XVIII. Transpiration Rate and Leaf Moisture Percentage in Different Forest Plantings

Species	Location	Period of measurement	Annual rainfall (mm)	Mean daily transpiration per tree (mg/g/h)		Mean daily transp. for season per tree (mg/g/h)	Mean daily leaf moisture per tree (%)		Mean daily leaf moisture for season per tree (mh/g/h)	Correlation co-efficient between transpiration and leaf moisture per-centage	
				maxi-mum	mini-mum		maxi-mum	mini-mum		r_1	r_2
Pinus halepensis	Eshtaol	16.4.62–21.3.63	486.0	72.95	24.16	45.52	55.00	45.50	51.71	0.31	1.63
Pinus halepensis	Hulda	1.5.62–7.4.63	407.0	94.83	29.50	45.06	54.70	49.10	51.30	0.00	−0.09
Pinus halepensis	Carmel	11.4.62–27.3.63	375.4	88.83	17.66	43.45	55.00	50.10	52.69	−0.12	0.15
Pinus halepensis	Galed	28.3.63–7.1.64	497.5	64.72	26.27	34.20	58.50	52.50	54.60	−0.28	0.40
Pinus halepensis	Burma route	4.4.63–12.12.63	280.8	82.75	30.38	41.02	44.40	51.40	53.00	0.08	0.20
Pinus halepensis	Eshtaol	4.1.65–18.5.65	703.5	123.99	76.34	96.57	56.49	51.42	53.53	−0.16	0.02
Pinus brutia	Eshtaol	17.1.65–20.5.65	703.5	85.97	46.26	63.79	58.98	55.53	57.23	0.63	0.41
Pinus halepensis	Lahav	16.8.65–28.7.66	244.0	103.16	59.90	83.45	59.39	50.58	55.16	−0.26	0.22
Pinus halepensis	Gevulot	6.7.65–25.8.66	129.4	87.50	42.60	60.91	59.00	49.90	53.81	0.12	−0.32
Eucalyptus camaldulensis	Gevulot	15.5.65–14.8.66	129.4	242.20	78.10	134.93	57.23	45.94	51.15	−0.24	0.48
Tamarix aphylla	Gevulot	1.9.65–26.8.66	129.4	142.10	56.40	113.05	68.61	59.22	62.78	0.31	0.60

1965), is impossible in the arid conditions after an equilibrium is established (June–July) because: 1. During the critical hot months soil is moister under the canopy of forest plantings, or even under the foliage of isolated trees, than in neighboring bare areas on identical topography and soil structure. In these circumstances there is not any upward or horizontal flux of water into the root zone, because more moisture is available with increasing proximity to the forest or an isolated tree.

While transpiration decreased to about 1/3–1/5 during the summer months August–September from its maximal rate in March–April, percentage leaf moisture

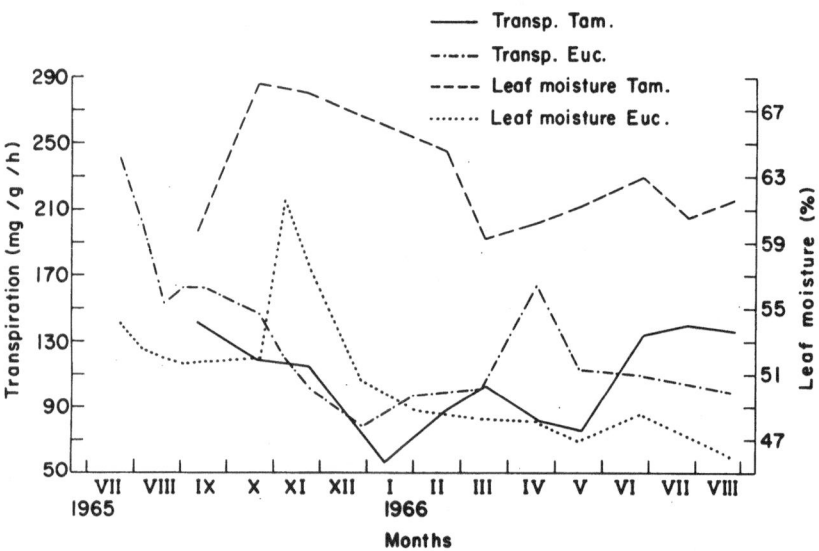

Fig. 45. Leaf moisture and transpiration rate in *Eucalyptus camaldulensis* and *Tamarix aphylla*, Gevulot.

in the studied tree species decreased by a maximum of 20 % only during the same period. For this reason there is a lack of parallelism between the decline in transpiration and the decline in moisture percentage in the leaves during the season of growth. This lack of correlation is in harmony with the former results concerning the relationship between transpiration and the water factor, which showed that this metabolic activity is governed by the same array of ecological factors as other life processes (GINDEL, 1967). Due to the small decrease of percentage leaf moisture, a proper cell turgor is maintained even during the critical months, and this enables transpiration to continue to some extent. The higher moisture content in *Tamarix aphylla* leaves even suffice for an abundant flowering during September.

Hence even in regions where there is only four of five inches of annual rainfall, trees can survive and foliage does not undergo any serious dehydration which may lead even to a temporary wilting. In order for the tree to survive, to maintain its green foliage during the critical months, and to continue its seasonal activities, quite a high degree of hydration must exist in the leaf. This hydration is not significantly different from that of tree species grown in more rainy climates (ACKLEY, 1954).

Lack of dehydration during the hottest months is reflected not only in the foliage but also in the roots, as revealed by LESHEM (1968). He studied Aleppo pine at two locations in the Judean Hills, in the neighborhood of our experimental plots. Figure 14 shows the variations in moisture percentage in the roots and within the soil in the root zone at the two locations, Panorama and Neve-Ilan. At the former, soil moisture diminished rapidly from January until June; from July until October a partial equilibrium was preserved. Concomitantly, moisture in the roots decreased to a minimum of 70%, and then started to rise from July into the rain period. At Neve-Ilan, soil moisture diminished from December until the end of June, and then began to rise slowly into the rainfall season. Root moisture decreased from January until April, and increased after June; its minimal value was about 74%. This would indicate that roots have even higher moisture contents during the critical months that those found in leaves.

Another feature associated with the regulation of foliage area due to water deficiency is clearly demonstrated in the acclimatized Eucalyptus species, which sheds a part of its older leaves during years of drought. The shedding of a portion of foliage, or the formation of dimorphic leaves in semi-shrubs have also been shown by VASSILJEV (1931) in *Smirnovia turkestana* in the Kara-Kum desert, and by ORSHAN (1954), in *Artemisia monosperma* and *Poterium spinosum*, which are indigenous to the Mediterranean maqui.

The reverse behavior has been noted in Aleppo pine. In this species, the new needles start to sprout in March, growing gradually by about 1–1.5 cm in length per month, and attaining maturity at the end of July. What is the value of an increase of foliage area (of about one-third) during the most critical months? There would appear to be no value to the plant in an increased metabolic activity and concomitant increased transpiration intensity. In any case, transpiration fell during the critical period to minimal values. Further study is needed to clarify whether the increase in foliage can improve the ability of the tree to absorb atmospheric moisture.

E. *Transpiration as a Function of Photosynthesis*

A parallelism between fluctuations in transpiration and in photosynthesis to a greater or lesser extent has been shown by SCHNEIDER *et al.* (1941), POLSTER (1950), MAKKINK (1957), BRUN (1963), SHIMSHI (1963), LANGE (1969), BRAVDO (1972) and others. SHIMSHI has shown that a reduction in soil humidity from 24 to 18%, a value still above the wilting point, affects both photosynthesis and transpiration, but to a larger extent photosynthesis. According to LEMON (1966), the complex photosynthetic process is more sensitive to protoplasmic water stress than is transpiration, and most means of reducing transpiration have an adverse

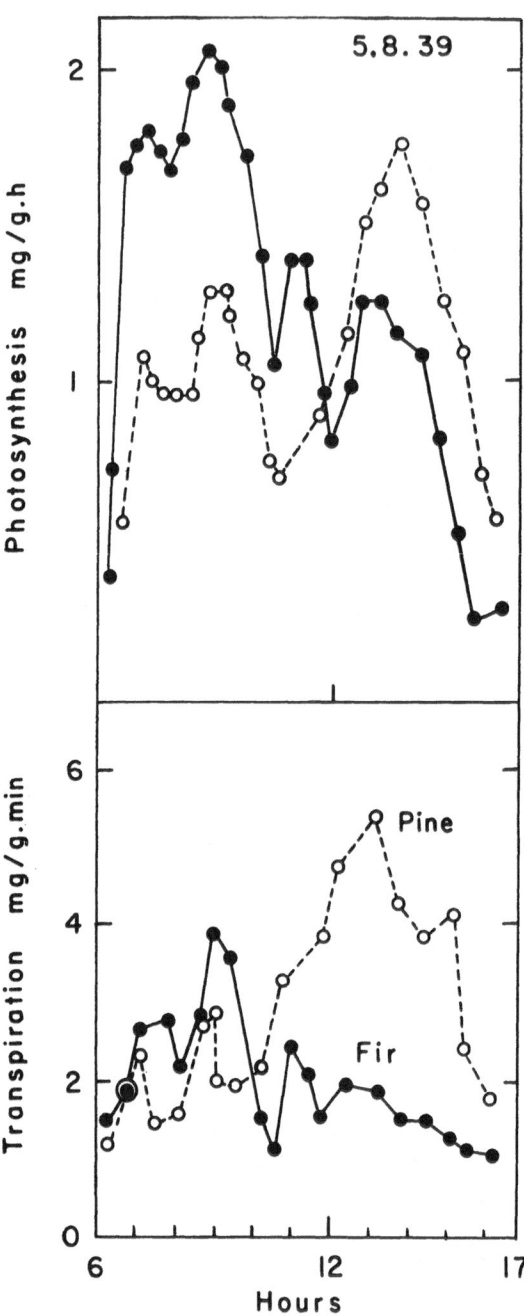

Fig. 46. Daily transpiration and photosynthesis in the fir and pine (POLSTER, 1950).

affect on photosynthesis. His conclusion that plants with a high net photosynthesis can also have a high rate of water loss through transpiration coincides with our results according to which transpiration rate is parallel to the rate of other physiological processes. POLSTER (1950) found a strict correlation in regard to the daily and seasonal fluctuations between transpiration and CO_2 uptake in some of the main forest trees grown in central Europe (Fig, 46). An increase in the water

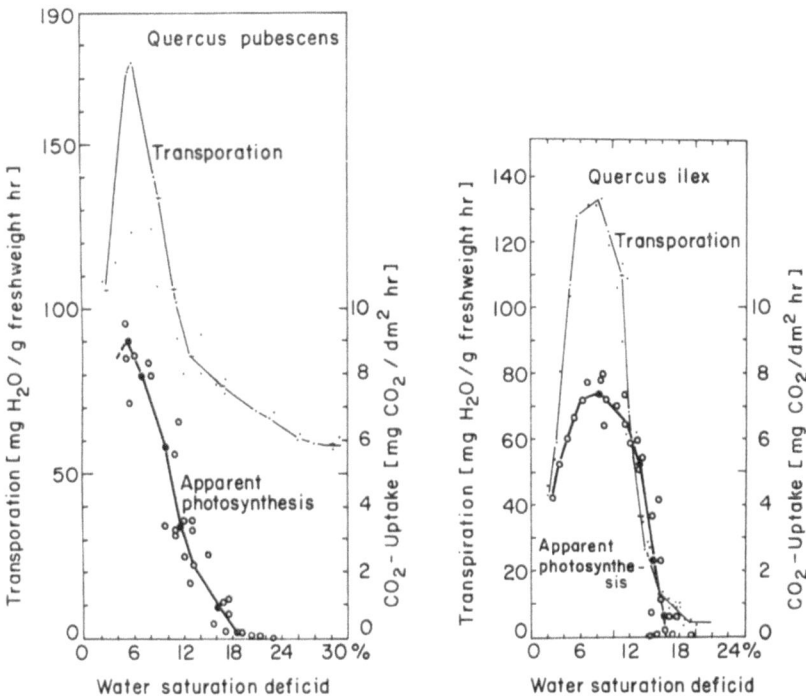

Fig. 47. Transpiration and photosynthesis as a function of soil moisture in two oak species. (According to LARCHER, 1960).

shortage greater than 5–6% affects both processes similarly in the case of *Quercus pubescens* and *Q. ilex*, (LARCHER, 1960) (Fig. 47); Samples were taken from the trees in the middle of July, 1959 (*Q. pubescens*) and in the middle of August (*Q. ilex*), when cambian activity is at a maximum. The decline of both of them at the same time, following the drying of the soil, has been shown by HEINICKE and CHILDERS (1936) (Fig. 48), and by SCHNEIDER and CHILDERS (1941). An obvious correlation of fluctuations in photosynthesis and transpiration in the alfalfa examined for 3 consecutive days (Fig. 49), has been shown by THOMAS and HILL

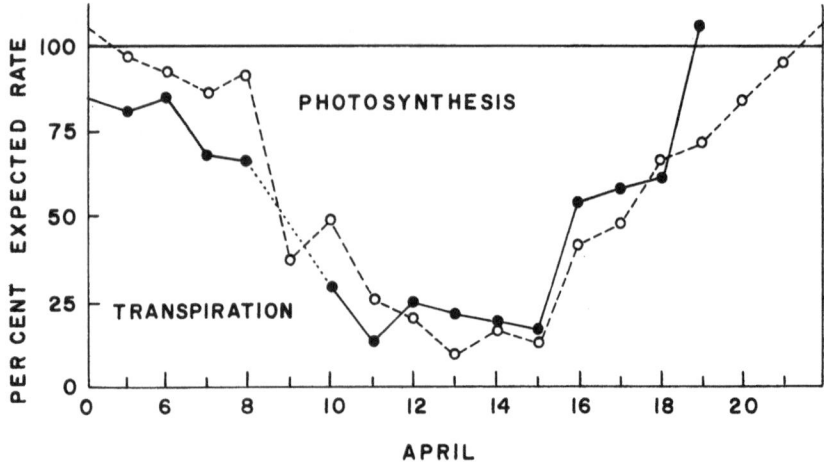

Fig. 48. The march of transpiration and photosynthesis in apple trees (According to HEINICKE and CHILDERS, 1936).

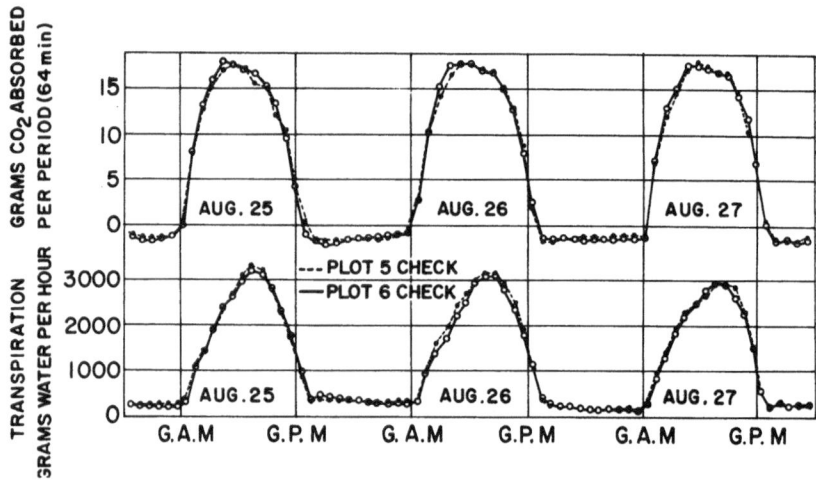

Fig. 49. Transpiration and photosynthesis (alfalfa) for 3 consecutive days (according to THOMAS and HILL, 1937).

(1937). These two processes showed maxima about noon, a drop toward the morning and evening hours.

POLSTER (1950) studied the transpiration and CO_2 uptake of six main tree species in central Europe and found conspicuous parallelism between these two processes. A similar strong parallelism between the two physiological processes has been

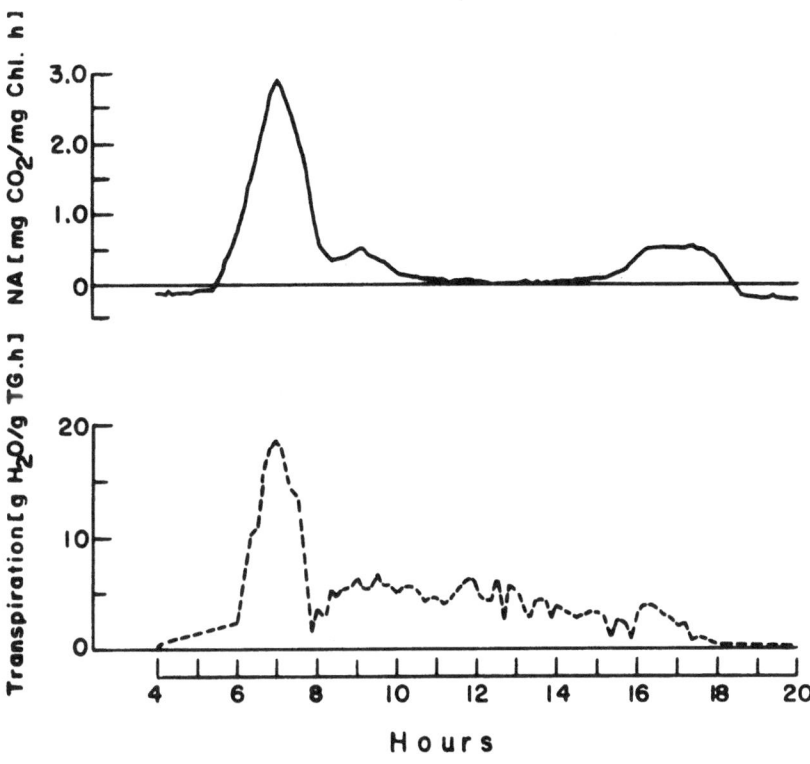

Fig. 50. Hourly fluctuations in net assimilation (NA) and transpiration. *Prunus armeniaca* (Avdat, Central Negev). LANGE, O. L., KOCH, W. and SCHULZE, E. D. (1968).

Average Daily Water Consumption and CO_2 Uptake and Photosynthesis

Species	Water Consumption g	CO_2 Uptake mg
Betula	9.50	22.85
Quercus	6.02	17.51
Fagus	4.83	11.13
Larex	3.34	7.93
Pinus	1.88	7.91
Pseudotsuga douglasii	1.38	5.45

stated by LANGE et al. (1967) in *Prunus armeniana* grown in extreme climatic conditions in the central Negev (Avdat). Yearly rainfall in this location ranges from 20–160 mm and maximal temperature rises to 45°C and more. As seen from Fig. 50, in the hottest months (August–September), in which the investigation was carried out, the daily fluctuations in CO_2 uptake and transpiration rate were identical.

F. *Transpiration as a Function of Cambium Activity*

A comparison of tracheid formation with temperature fluctuations shows that in normal years the cambium is inactive for most of the rainy season in spite of available soil moisture (GINDEL, 1967). This inactivity occurs because of the drop of temperature to its minimum values. In years of drought (Table XIX), when the number of rainy days is scarce and temperatures are relatively high, the transpiration rate is also conspicuous during the rainy season.

Table XIX. Rainfall and the Index of Aridity at Ben Shemen

Year	Rainfall (mm)	No. of days of rain	Index of aridity
1945	649.8	80	30.7
1946	631.4	49	16.9
1947	376.0	45	9.8

Toward the end of the rainy season or immediately afterwards, when the absolute minimum temperature exceeds 6.0–7.0°C, cambium activity in Aleppo pine is renewed. Cambium began to be active at Ben-Shemen in February, 1946, and in January, 1947, with an average minimum temperature of 7.2°C and 8.1°C.

A glance at the layer of xylem cells formed in a dominant tree during a year of average rainfall (Fig. 51) reveals that:

1. During March or April, depending on the elevation and temperature, the first tracheids appear; they have thin walls and wide lumina;
2. In mid-May, the tracheid walls commence to thicken and the lumen narrows; this narrowing continues until the beginning or the middle of July;
3. From July or the beginning of August until the end of August or the beginning of September, tracheids are formed with very thick walls and extremely narrow lumina. These first and third tracheid layers in Aleppo pine resemble the early and late wood of pines growing in temperate or cold regions. In contrast, the second layer is peculiar to Aleppo pine, and it reflects the influence of the subtropical climate on the pattern of the tracheids;

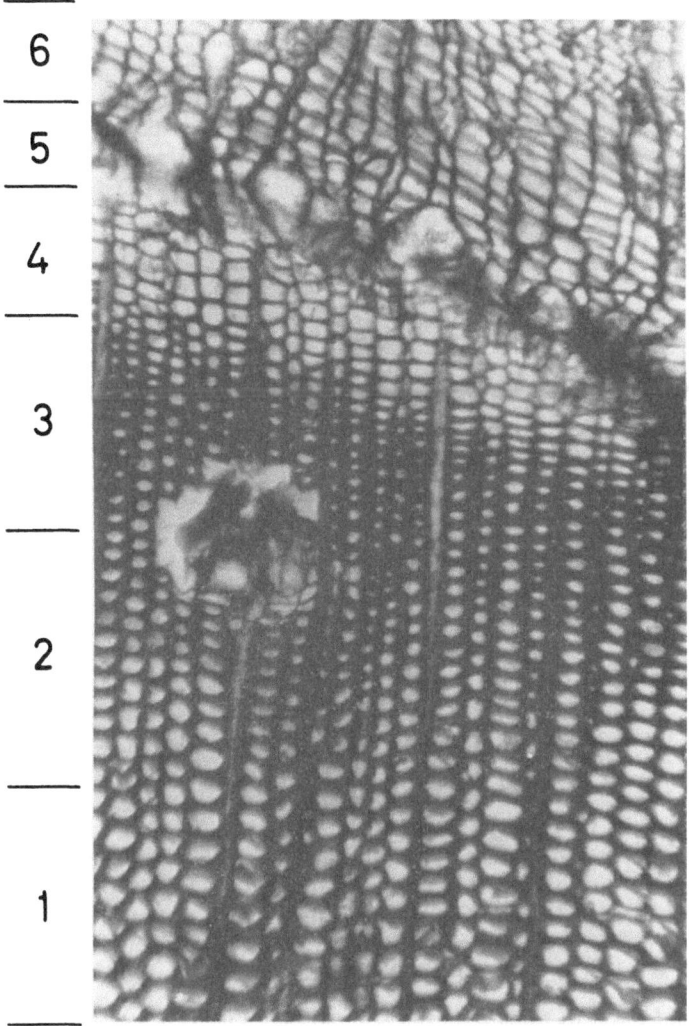

6

5

4

3

2

1

Fig. 51. The pattern of the tracheids in annual ring in the trunk of an Aleppo pine: 1. Spring wood. 2. Intermediate wood. 3. Late wood. 4. Autumn wood. 5. Cambium. 6. Phloem. (Mt. Carmel).

4. At the end of August, or during September, cambial activity often almost ceases, but it may be renewed after the first rains (October or November) if the temperature is suitable. This phenomenon is especially common at sites of low elevation (Ben-Shemen, Hulda, Carmel). Tracheids formed under such conditions partly resemble the second layer and partly the third, and they give the

101

impression of an annual ring; however, it is really a "false ring". A similar phenomenon is known in *Pinus insignis* on Mt. Carmel in California (GINDEL, 1944).

5. During the wettest months (December–February) with, generally, the lowest temperatures, cambial activity ceases completely. On the other hand, in low-lying regions (as in the sands at Rehovot) sporadic tracheid formation often continues throughout the rainy season (Table XX). This table shows the fraction of tracheids formed monthly. In a year of normal rainfall, cambial activity continues for 8–9 months at Ben-Shemen, while in drought years, as in 1947, 1957, 1961 and 1963, the formation of tracheids stopped after 6–7 months.

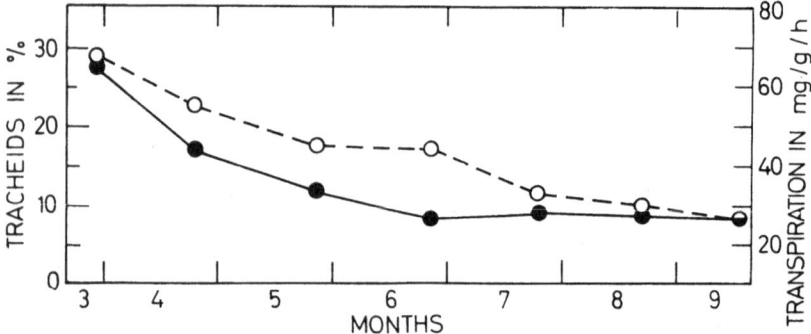

Fig. 52. The fluctuations in the rate of transpiration per tree (broken line) and percentage of tracheids (solid line) formed monthly in Aleppo pine forest (Efraim Hills).

6. Almost half of the tracheids were formed from 15.3 to 15.5, and over 90% had appeared by mid-August, whereas this fraction was formed by mid-July in the years of drought. In such years, there is almost no formation of the intermediate tracheid layer, and only early and late wood can be distinguished. These appear similar to the two layers found in suppressed trees.

During the years 1961/62 the rates of transpiration and percentages of tracheids formed monthly were calculated from data from 6 dominant trees, in a 12-year-old forest in the Hills of Efraim grown in rendzina soils (Fig. 52). The forest was on a northerly slope of about 30% and the depth of soil above bedrock was 90–120 cm.

The new tracheids which began to appear in the trunk before the 24th of March amounted to 28% of the total number produced during the growing season. By the end of April, another 17% had been added, and, by mid-July, about 70% of the total had appeared. During the later summer months, until mid-September, about 10% was added monthly in the two forests. At the end of March, heightened cambial activity was accompanied by a higher rate of transpiration (close to 65 mg/g/h). At the end of May, this rate was still appreciable (about 33.3 mg/g/h), and it dropped to 25.8–26.6 mg/g/h throughout the rest of the year.

Table XX. Monthly Tracheid Formation in % at Maaleh Hachamisha (a), Ben-Shemen (b), and Rehovot (c)

Year	Locality											
1945	(dates)	15.III	15.IV	13.V	15.VI	15.VII	16.VIII	15.IX	15.X	14.XI		
	(a)	—	24.5	41.1	56.0	67.0	76.0	86.0	93.5	100		
	(b)	16.2	30.5	45.5	50.00	66.3	84.0	89.0	98.0	100		
1946	(dates)	26.II	15.III	14.IV	14.V	1.VII	1.VIII	1.IX	15.XI			
	(a)	3.0	21.5	42.5	69.0	80.0	83.0	94.0	100			
	(b)	6.2	18.1	33.2	45.3	66.6	87.0	94.8	100			
1947	(dates)	16.I	5.II	10.IV	8.V	17.VI	13.VII	15.VIII				
	(a)	5.4	14.2	30.0	49.0	87.0	92.0	100				
	(b)	4.8	12.0	32.1	63.3	54.6	96.0	100				
1954	(dates)	1.II	7.III	1.IV	4.V	2.VI	1.VII	1.VIII	30.VIII	29.IX	1.XII	3.1.55
	(c)	7.7	17.9	27.8	48.5	72.5	82.5	89.1	90.0	94.0	2.2	3.9

A similar parallelism between the fluctuations in the rate of transpiration and the intensity of cambium activity, has been noted in the main tree species of the Mediterranean Maqui and in the following introduced species: *Pinus pinea, P. canariensis, Eucalyptus camaldulensis.*

G. *Transpiration and Growth*

The relationships of transpiration and growth were examined by comparing the tree-ring area of the tracheids in the trunks formed during the growing season with the mean hourly transpiration for the season of growth The study was conducted by grouping 6 dominant trees in an experimental plot in 4 different locations. In each plot the trees were divided into two groups: one with higher and the second three with lower rates of transpiration. Table XXI shows that a larger, year-ring area accompanies more intense transpiration, and vice versa. This phenomenon appeared in all experimental plots during the six years of studies, during which time the amount and distribution of rainfall differed.

Table XXI

The Correlation between Rate of Transpiration per Tree and Tree-ring Area

Sample plot	Year of testing	Three trees of high transpiration rate		Three trees in low transpiration rate	
		Transpiration in mg/g/h	Tree-ring area in mm²	Transpiration in mg/g/h	Tree-ring area in mm²
Eshtaol	1962	125.0	156.0	82.7	89.0
Hulda	1962	117.0	145.6	77.3	81.6
Ataturk Forest	1962	137.0	216.7	65.8	116.3
Burma Road	1963	121.3	83.1	109.0	65.0

A further step taken in the establishment of the interdependence of the two processes was to compare the rates of transpiration in dominant and suppressed trees growing close together in the same experimental plot. This experiment was performed in 1958–1961 for the 6 species shown in Table XXII. These were growing in the Rehovot Arboretum in sandy, unirrigated soil. The transpiration rate was calculated for 3 dominant and 3 suppressed trees. Transpiration was greater for the dominant than for the suppressed trees in all 6 species. In *Eucalyptus gomphocephala* the difference was greater than 50% during August. In Aleppo pine, the difference on January 21, 1961, for suppressed trees was only 65% of that found in the dominants. In *Acacia accuminata* the maximum difference was on October 25, 1959, when the transpiration of the suppressed tree was 10% of that of the dominant. The differences in the rates between dominant and suppressed trees increased as soil moisture became more critical.

Identical results were obtained in a 30-year-old Aleppo pine plantation in the Judean Hills (Hulda). Transpiration was measured in three dominant and three

Table XXII

Mean Hourly Transpiration in mg/g/h (Calculated for an 8-hour Day, 8 a.m. to 4 p.m.)

Date	Dominant	Suppressed	Suppressed/ dominant (in %)
	Eucalyptus gomphocephala		
2.4.59	205.2	114.8	56.0
30.6.59	179.4	159.6	88.9
6.8.59	286.7	139.2	48.5
28.3.60	122.7	83.1	67.7
2.8.60	214.7	147.0	68.5
	Eucalyptus poleanthemos		
2.4.59	228.2	224.5	98.4
21.5.59	367.3	333.2	90.7
30.6.59	296.2	261.5	88.3
6.8.59	181.9	144.6	79.5
25.10.59	84.9	65.4	77.0
28.3.60	185.0	165.7	89.6
	Eucalyptus preissiana		
1.5.58	136.3	96.8	71.0
20.10.58	43.7	27.0	62.0
17.3.59	109.9	87.6	80.0
	Eucalyptus torquata		
1.5.58	218.0	209.8	96.0
17.3.59	109.1	56.8	52.0
	Quercus ilex		
20.10.58	70.5	54.4	77.0
17.3.58	157.0	110.6	70.4
	Acacia accuminata		
20.10.58	132.0	113.4	85.1

suppressed trees (Fig. 53). The diameter of the dominant trees varied from 19.5–34.0 cm, and their height from 5.5–9.8 meters, whereas for the suppressed trees the diameter was 8–9 cm, and the height 3.5–3.6 meters. Figure 53 illustrates the course of the daily transpiration averaged for the 3 dominant and 3 suppressed trees in 4 different months.

The conspicuous differences in the two groups of trees is a further evidence of the correlation between growth and transpiration. More intensive growth of the dominants is accompanied by a higher rate of transpiration.

The intention in these experiments was not to prove that the amount of water transpired by dominant trees exceeds that of suppressed trees, since this is obvious

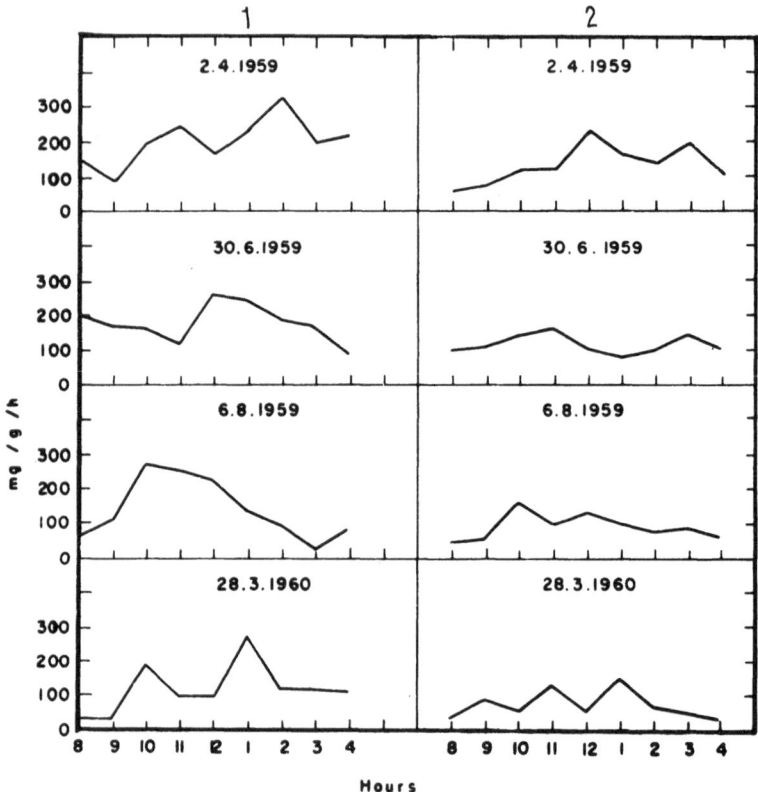

Fig. 53. *Pinus halepensis*. Differences in intensity of transpiration (the sum of 3 trees) in dominant (1) and suppressed (2) trees in a 30-year-old forest (Hulda).

and is due to the much larger number of leaves on dominants. The aim was to compare an equal number of leaves in the two cases, and to prove that one gram of green matter from the dominant trees transpires more water per hour than it does from the suppressed ones.

There is often a marked correlation of a similar type between these two physio-logical processes when they are compared in species in which the rate of growth is different. Two pine species were examined in 1964–1965: *Pinus halepensis* Mill and *P. brutia* Elw. and Men. The latter species was introduced into Israel from Cyprus;

its growth is relatively slower than in the former species. Two forests of these two species, in the Hills, were examined. The trees were 14 years old; they grew near one another along a rocky southwest slope in shallow terra rossa soil. Table XXIII shows the mean hourly transpiration rate for 6 trees of each species, and the average

Table XXIII.

The Mean Transpiration-rate and Average Trunk Volume in Aleppo Pine and Brutia Pine

Species	Transpiration mg/g/h	Average diam. (inches)	Average height (feet)	Volume (c.f.)
P. halepensis	100.2	6.8	32.4	0.47
P. brutia	64.7	6.4	29.5	0.35

trunk-volume. The transpiration rate is 38.5% greater in the Aleppo pine and its growth in volume is 25.6% more.

To what extent tree dimensions affect transpiration was shown in an experiment carried out in the desert (Gevulot) (GINDEL, 1968) in three neighboring forest plantations. Each plantation was of a single species: *Tamarix aphylla*, indigenous in the desert, *Pinus halepensis*, indigenous to the Mediterranean maqui, and *Eucalyptus camaldulensis* indigenous to the subtropical zone of Australia (Table XXIV). From each planting, six dominant trees were sampled, and the two pairs

Table XXIV. The Correlation between Mean Transpiration Rate and Tree Development

Species	Tree number	Transpiration in mg/g/h	Height in m	D.B.H. in cm
Eucalyptus	2	125.0	5.5	12.4
	6	112.7	7.0	14.0
	3	137.8	7.5	15.0
	4	136.8	7.8	16.0
Tamarix	2	115.8	6.0	25.6
	3	107.3	4.0	17.4
	4	110.4	4.8	12.7
	5	119.4	6.5	19.0
Pinus (Gevulot)	2	63.6	6.0	11.4
	6	67.6	7.0	9.8
	3	56.5	5.0	8.4
	4	50.8	5.0	9.2
Pinus (Lahav)	1	99.3	9.5	12.4
	3	79.8	7.5	12.9
	4	74.7	8.5	11.0
	6	87.6	8.0	12.0

which had extreme values of transpiration throughout the period of the experiment were compared. The table shows that there is a relation between tree dimensions in all three species and their transpiration rates. The larger trees, both in height and girth, have higher rates of transpiration in all three species studied.

A survey of the literature revealed that results of many workers show transpiration to be correlated with growth in mesophytic climates too. According to POSTER (1950), there is a parallelism between transpiration and photosynthesis on the one hand, and the amount of woody tissue formed during a season of growth on the

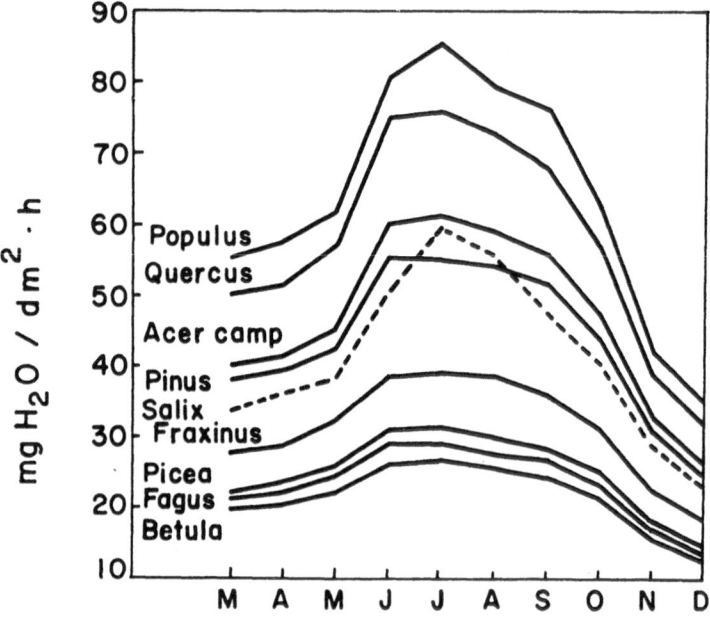

Fig. 54. The fluctuation in transpiration rate during the different months of the year in 9 European species grown in mesic climatic conditions (GEURTEN, 1950).

other. For rapidly growing trees, a high rate of photosynthesis is accompanied generally by a high rate of transpiration. *Pseudotsuga douglassi*, for example, exhibits strong photosynthesis and active transpiration, while *Pinus sylvestris* shows weaker activity for both these processes.

In the Death Valley of California, the hottest and driest region of the United States, where yearly rainfall is 20–80 mm, the annual temperature extremes are −2° to +50°C, accompanied by persistent and often hot winds, similar results have been obtained by STARK (1969). In three dominant woody species (*Atriplex hymenelytra*, *Larrea divaricata* and *Penecephyllum schottii*), transpiration does appear to increase during the period of active growth. It follows that when the plant has to withstand drought and to save water the rate of growth is hampered or even entirely interrupted.

The parallelism between rate of growth and rate of transpiration is reflected also in the quantity of ash. According to DE WITT (1958) and HANKS et al. (1967), production and transpiration are directly related for sorghum, wheat, and other agricultural crops.

Table XXV and Fig. 54 which show transpiration rate of the main forest tree species in Europe, are further evidence that transpiration is highest during the main growing season (May–June until September–October). Identical results were obtained by CIAMPI (1954), for *Abies alba* and by GINDEL (1959), for the same species in the Appenins (Vallombrosa).

Table XXV. Rate of Transpiration of European Trees (POLSTER, 1950)

Species	Daily transpiration from two green leaves	
	19.7–6.8	21.9–28.9
European Larch	4.6	2.2
Norway Spruce	1.0	0.5
Scotch Pine	1.9	0.7
White Pine	2.1	0.7
European Beech	4.1	2.3

According to SCHUBERT (1940), the rate of transpiration is 2–3 times greater during the period of intense growth than toward the end of the season.

In the temperate zone, the same temperature (20–30°C) which leads to a high transpiration rate is also accompanied by high rates of photosynthesis, gaseous exchange through open stomata, absorption of salts in solution from the soil, transportation of carbohydrates produced in chloroplast-containing tissues, enzyme activity, and the formation of pigments, glycosides, amides, and amino acids used for the formation of new protoplasm in the cambium.

THE XEROMORPHIC PROPERTIES OF THE LEAF AND THEIR RELATIONSHIP TO THE PROCESS OF TRANSPIRATION

Erroneous interpretations of the process of transpiration led some workers to an unrealistic interpretation of the morpho-anatomical structures of the leaf. MAXIMOV (1929) referred to the leaf as an unsuccessful organ open to the whims of its surroundings. He claimed that rate of transpiration increases as the climate becomes drier. In the opinion of STALFELT (1956), the form of the leaf is not well suited to the water-economy of the plant, and it does not prevent water-shortages from being imposed on the plant by environmental factors. MILLER (1938) claimed that the structure of the leaf is not suited to the task of maintaining its water content, and that the perpetual loss of water endangers the life of the plant because moisture is steadily exhausted in direct proportion to atmospheric demands.

Later a third concept was developed, i.e., that "transpiration utilizes excess heat and helps to prevent plants from becoming overheated" (THORNTHWAITE and MATHER, 1951). GATES (1968), even assures that "if the plant physiologist ever has had a doubt concerning the value to the plant of transpirational cooling as it affects leaf temperature, this gross misconception should be dispelled now once and for all."

The adjustments of the leaf to arid climates varies with species and phytogeographical location. The most striking are:

1. A specific pattern of stomata (analyzed in the next paragraph).
2. Extremely fleshy succulents, with sword-shaped leaves and spines at the tips or edges (AGAVE, ALOE, FOURCROYA, etc.).
3. Fleshy leaves of the type of *Mesembryanthemum*, which resemble flint stones, are found in the semi-arid and arid zones of Africa.
4. Development of subterranean leaves, as in the succulent *Lithops salicoal.*
5. Deposition of a silver-gray powder(*Eucalyptus*): In the rainy, cool season the powder often disappears and the leaves revert to their normal green color (e.g., *Eucalyptus pulverulenta*).
6. Greater vein density.
7. A gray felt covering.
8. Salt-crystals over a large part of the leaf (*Tamarix sp*).
9. Bimorphism: development of larger leaves during the wet season, and fewer smaller ones in the dry season, in order to reduce physiological activity (*Poterium spinosum, Salvia triloba, Thymus capitatus, Artemisia monosperma*, and *Varthemia iphionoides*).
10. Folding-over of leaf-pairs during intense light.

11. Leafing during the rains, and leaf-fall in the dry season, when the green branches assume the physiological activity previously maintained by the leaves. Such plants are very common in the African, American, Australian, and Asian deserts. For example: *Calligonum commosum, Fouquieria sp., Genista sp., Ochrodenus baccatus, Retama sp., Zygophyllum dumosum; Capparis,* and *Croton floribundus.*
12. Brachyblasts in place of leaves, for example, in *Atriplex sp., Heliotropum rotundifolium,* and *Suaeda asphaltica.*
13. Stomatal apertures sunk into the epidermis, in hollows with wax, and often protected by crypts.
14. Spiniform leaves (*Mulinum*) or squasiform imbricated (*Lepidophyllum*). Leaves set "edge on" to avoid direct illumination ("Caatinga" shrubs) or the frequent presence of spines.
15. Palisade cells arranged towards both surfaces of the leaves as closely knit layers with closer packing at the upper, more exposed surface.
16. Diminution of the extent of the spongy tissue, with the intercellular spaces forming a close-packed mesophyll of minimum volume, in contrast to mesomorphic leaves where such tissue forms 1/3 or more of the volume of the leaf. Also, the presence of hypodermal sclerenchyma. All these modifications increase the quantity of ash.
17. Small intercellular spaces, and diminution of the epidermal cells and leaf area.
18. Deposition of suberin and cutin in the cuticle to prevent cuticular transpiration. Essential oils secreted by leaves of many xerophytes may bring about a diminution in the rate of water-vapor diffusion. The leaf is often filled with resin or wax.
19. Mucilaginous materials which increase the osmotic pressure of the cell fluids and heighten the power of absorption (e.g., tannins, gums, and oils in Eucalyptus and many other species, rubber in *Parthenium argentatum,* and milky juice in the *Asclepiadaceae*). Mucilaginous juices are often found in the thick fleshy stems of the Cactaceae. These substances, together with the hypodermal sclerenchyma, increase the insulation of the leaf from extreme climatic conditions.
20. Epidermis covered with colenchyma or hairs of various kinds: some glandular hairs contain chlorophyll and protoplasm and take part in metabolism, others lack these materials. Although the functions of these hairs have not been sufficiently defined, various investigations have shown that the physiological activities of the plant are disturbed if they are removed.

A further adaptation for survival under arid conditions is the occurrence of groups of plants which cover the soil surface by the formation of a more or less closed semi-circular canopy of foliage. Such groups of plants are often composed of various species, as may be seen along rocky hillsides or in arid sand-dunes in sub-tropical regions. A semi-circular foliage protects an area almost double its own base from physical evaporation.

Another xeromorphic adaptation is apparent in the genus Eucalyptus. In the early years, large juvenile leaves appear. After the plant has become established and grown further, sometimes as early as after 2 or 3 years, intermediate-sized leaves appear which differ in size, toughness, and other anatomical details. After a few more years of normal development, adult leaves of a xeromorphic structure appear at the top of the tree and develop throughout the rest of its life. The juvenile and intermediate-sized leaves on the lower part of the stem gradually fall, leaving only the adult ones. In some kinds of Eucalyptus the intermediate leaves do not appear, but there is a direct transition from juvenile to adult leaves.

The juvenile leaves are arranged laterally, while the adult ones are set at an angle to the direction of the light. This change in direction results from responses of the leaf blade and the structure of its petiole when climatic conditions worsen, as in the desert, where Eucalyptus assumes a shrub-like form (Marlock, Mallee) of a xeromorphic foliage without the bi- or trimorphism of the leaf.

In the detailed xero-morpho-anatomical structure of the leaf, the stomata play the main role in the process of gaseous exchange for photosynthetic activity.

A survey of the indigenous and acclimatized tree species in Israel revealed the unreality of the accepted opinion that woody plants have stomata on the lower epidermis only. This ideas does not coincide with the situation found in many indigenous and acclimatized woody species (GINDEL, 1969). In the majority of them, stomata are present on the lower and upper sides of the blade. Moreover, such species grow successfully in the most extreme conditions, even along the rocky slopes of the Dead Sea where yearly rainfall is only two inches; as for instance, *Acacia spirocarpa, Anabasis articulata, Ochrademus baccatus, Suaeda monoica* and others. The same holds for the acclimatized woody species growing there in irrigated gardens, such as *Cassia sp., Parkinsonia aculeata, Prosopis sp.*, and others.

From an investigation carried out in this respect on indigenous and acclimatized tree species the following facts have been revealed:

1. Stomata are found on both sides of the blade in 21 species presented in Tables XXVI and XXVII. Likewise in some of these species stomatal density is significantly higher on the upper epidermis than on the lower. Moreover, some of these species are known in Israel as prominent xerophytes grown in poor soil and under extreme climatic conditions. Some even succeed in growing in the desert, where subterranean water is far below the root zone. These species maintain their green foliage during the critical months of the dry season (July–October), and some survive in the desert even during years of drought, without any artificial irrigation with only 80 mm of rainfall, and in the subtropical zone with 200–300 mm of rain.

2. Eight species have more stomata per unit area on the upper epidermis than on the lower, the differences ranging between 5 and 63%, depending on species. Sixteen species often have two or three times more stomata on the lower epidermis than on the upper. They are distinguished by very prominent xerophytic properties and some of them, e.g., *Acacia cyanophylla* and *Schinus molle*, grow in the semi-desert area or in the desert.

3. In some of the exotics listed in Tables XXVI and XXVII and in others studied but not presented here (such as *Acacia ciliata, A. pendula, A. saligna, Eucalyptus*

forrestiana, and *E. occidentalis*), in which some of the leaves are oriented vertically on the tree, the differences in the number of stomata per unit area in the upper and lower epidermis are small and insignificant. These differences are smaller than those in the leaves of the same plant which are oriented horizontally.

Table XXVI. Density and Length of Stomata in Various Tree Species Grown without Irrigation in Poor Sands at Rehobot, Israel

Species	Number mm²		Length (μ)	
	Upper epidermis	Lower epidermis	Upper epidermis	Lower epidermis
Indigenous				
Balanites aegyptiaca	17	46	23	25
Pistacia atlantica	148	386	22	26
Introduced				
Acacia longifolia	153	104	21	22
Callistemon laurifolia	403	361	29	26
Callistemon rigidus	253	259	30	35
Eucalyptus albens	152	128	32	30
Eucalyptus blakelyi	323	354	19	19
Eucalyptus cinerea	209	233	20	22
Eucalyptus ficifolia	167	328	27	22
Eucalyptus maculata	147	242	41	39
Eucalyptus maideni	33	36	67	58
Eucalyptus melanophloia	111	184	21	30
Eucalyptus melliodora	161	164	27	23
Eucalyptus sideroxylon	340	284	18	18
Eucalyptus staigeriana	243	292	40	37
Eucalyptus tesselaris	343	291	36	40
Melaleuca styphelioides	599	367	21	23
Myoporum serratum	99	200	30	30

The lowest stomatal density was found in the local *Balanites aegyptiaca* and the acclimatized *Eucalyptus maideni*. Stomata of the latter species are the longest. *Melaleuca armilaris*, whose leaf area is the smallest, has the highest stomatal density of the 32 species reported.

4. Irrigated trees of half of the 14 species presented in Table XXVII have fewer stomata per square millimeter on the upper as well as on the lower leaf epidermis than nonirrigated trees (differences are significant at the 1 % level). The differences in stomatal density between irrigated and nonirrigated trees showed a positive dependence upon the amount of water supplied to the irrigated trees and the rigor of the conditions under which the nonirrigated ones were growing.

114

Table XXVII. Number and Length of Stomata and Mean Leaf Area in Irrigated and Nonirrigated Tree Species

Species	Age of trees (yr.)	Location	Number of stomata per mm				Length of stomata in microns				Mean-leaf area (cm²)	
			Irrigated Epidermis		Nonirrigated Epidermis		Irrigated Epidermis		Nonirrigated Epidermis		Irrigated	Non-irrigated
			Upper	Lower	Upper	Lower	Upper	Lower	Upper	Lower		
Indigenous												
Ceratonia siliqua	11	Judean Hills	—	289	—	345*	—	—	—	—	12.1	8.4*
Citrus paradisi	9	Gilat	—	713	—	780*	—	14	—	14*	—	—
Citrus sinensis	9	Rehovot	—	580	—	682*	—	22	—	18*	24.9	29.0†
Phillyrea media	14	Mt. Carmel (North)	—	472	—	513*	—	18	—	21*	0.96	1.1†
Pistacia lentiscus	18	Mt. Carmel (South)	—	280	—	299*	—	—	—	—	0.90	0.70*
Prunus amyadalus	12	Negev desert	—	327	—	360*	—	30	—	29*	12.2	12.4†
Introduced												
Acacia cyanophylla	8	Judean Hills	322	307	421	398*	26	27	23	26*	9.2	13.5†
Eucalyptus camaldulensis	12	Gilat	81	119	89	121*	37	36	32	37*	18.5	22.0†
Eucalyptus citriodora	18	Rehovot	256	248	306	288*	26	24	25	24*	17.2	18.4†
Melia azedarach	8	Judean Hills	—	374	—	423*	—	19	—	16*	5.8	6.1†
Pittosporum tobira	6	Rehovot	—	368	—	415*	—	21	—	20*	12.0	9.3†
Pittosporum undulatum	6	Rehovot	—	339	—	557*	—	17	—	18*	16.0	19.3†
Pyrus malus	10	Jerusalem	—	444	—	534*	—	24	—	22*	16.2	11.4*
Schinus molle	9	Judean Hills	178	141	206	143*	26	25	25	23*	1.2	1.4†

* Difference between irrigated and nonirrigated significant at the 1% level.   † not significant.

Under the extreme climatic conditions of the desert, *Acacia cyanophylla*, when grown on a sloping area in deep loess, had 15.5% more stomata per unit area than trees grown close by but in a depression, where gravitational water concentrated during the rains, increasing two-fold and more the available soil moisture percentage. In *Ceratonia siliqua*, also, site quality affected stomatal density: on the rocky calcareous slope where there was only a thin soil layer of 25–30 cm, a significantly higher stomatal density of 16% was found in comparison to the trees grown on the same slope but where the soil was deeper. In the two Citrus species, the difference in the quantity of irrigation had a similar influence on the density and size of stomata.

Mean leaf area of measured leaves was larger in eight species of nonirrigated plants and in five species of irrigated (Table XXVII). Except in the cases of *Ceratonia siliqua* and *Pyrus malus*, these differences were not significant. An increase in stomatal density caused by water deficiency and poorer ecological conditions was not accompanied by a decrease in leaf area in 61% of the species tested.

These results and those obtained earlier (GINDEL, 1968–1970), indicate that further study is needed to clarify to what extent augmented stomatal density and the presence of stomata on the upper epidermis in woody xerophytes grown in the driest climate and poorest soil serve to increase the ability of the tree to absorb atmospheric moisture, when moisture tension in the soil and within the plant reach their highest values.

TRANSPIRATION SUPPRESSANTS

The identification of the process of transpiration with physical evaporation, and the foliage as the evaporator, led to the idea of covering a part of the stomata with chemicals or plastic films. This approach follows from the success in reducing the evaporation of water from reservoirs by covering their surfaces with mono-molecular films of fatty alcohols.

The identification of water lakes and dams with the foliage of plants is in har-mony with the physico-meteorological approach that the plant and the environ-ment are merely a series of conductors, and that through each conductor the same amount of water flows. Theoretically, the decreasing transpiration is only a matter of increasing the diffusive resistance in the water vapor pathway.

The use of transpiration suppressants stands in violation of our results as far as natural woody plants are concerned. Transpiration is a link of metabolism and is closely correlated and harmonized with the micro-climatic and edaphic conditions. An interference with the transpiration-metabolism link affects the different physio-logical processes simultaneously. Indeed, from the results of experiments with transpiration suppressants carried out by different workers, it follows that the diminution of transpiration was achieved at the cost of a decline in growth or of photosynthetic activity (WAGGONER, 1965), or even deformation of the plants (ANGUS and BIELORAI, 1965). By using silicon 40 to 50, transpiration dropped to 10.3–66.4% with a simultaneous diminution of dry matter by 63%. BARR (1945) advised that while the blockage of the leaf stomata by spray does, in fact, reduce the rate of transpiration it also causes a decline in the other biosynthetic activities. Stomatal closure induced by phenylmercuric acetate reduced transpiration and photosynthesis particularly when available moisture was decreased (SHIMSHI, 1963). PLAUT et al. (1968), by using various chemicals, among others, trimethyl-ammonium chloride, found a decrease in the top:root ratio in all treatments.

A drastic decrease in absolute photosynthesis was observed by GAASTRA (1959). A spray of 300 ppm phenylmercuric acetate on the needles of *Pinus resinosa* Ait. by WAGGONER and BRAWDO (1967), led to a smaller depletion of 20–28 mm of soil moisture with simultaneous decrease in growth of the bole by 15%. Photosyn-thesis decreased more than transpiration when cotton plants were sprayed with different suppressants (SLATYER and BIERHUIZEN, 1964). NEWMAN and KRAMER (1966), noted that roots of intact bean plants were killed by a 1-hour immersion in 10^{-3} MDSA.

KOZLOWSKI and CLAUSEN (1967), did not find significant improvement of the internal water balance of several woody species after they were sprayed with alkenyl or decenylsuccinic acid, and toxicity to roots often followed treatment. The author concluded that users of the transpiration suppressants appeared to be over-optimistic with respect to their practicality in inducing drought hardiness.

WAGGONER (1965) concluded his survey on transpiration suppressants by asserting that it seems a nearly impossible specification to find a synthetic film that could be ideally resistant to water and completely permeable to carbon dioxide.

BRAVDO (1972) investigated the influence of some transpiration suppressants in Citrus and grapevine seedlings. The author found that all the chemicals used reduced photosynthesis. The ratio of photosynthesis to transpiration did not increase in Citrus; some increments of this ratio have been noted in regard to the grapevine seedlings but the author concluded that by increasing water use, efficiency was always followed by inhibition of photosynthesis which, of course, is generally accompanied by loss of yield.

So far as it concerns woody plants grown in arid conditions, available soil water and favorable climatic factors control transpiration and hence, growth. Generally, the actual transpiration of the leaf is always below the maximum possible transpiration from the same leaf under the relevant meteorological conditions. As it was shown in the previous chapter that a smaller leaf, but a xeromorphic one, whose transpiration rate is scarce, when grown in a soil moisture deficiency, raises its transpiration rate immediately after soil moisture was replenished. Moreover, its rate of transpiration equalled or was even higher than the neighboring plants growing steadily in field capacity and of a larger mean leaf area by 40% and more. Therefore, artificial closure of the stomates impedes normal life processes of plants.

XEROPHYTISM

From the preceding paragraphs concerning the different ecophysiological life phenomena of the xerophytes we are able to sum up the basic properties of xerophytism.

Since SCHOUW, 1822, when the term xerophytism first appeared in the professional literature, it has taken many different meanings.

One should distinguish carefully between climatic and edaphic aridity where both types are of concern, as is the case in this investigation. In spite of the most extreme climatic factors found on this globe, plants will grow if edaphic water is available. Some species, however, are independent of soil moisture, and have their roots kept in the air to collect atmospheric moisture.

In arid regions where rainfall is only 3–6 inches annually, xeromorphism is diminished and growth rate conspicuously accelerated when in addition to the scarce rainwater, gravitational water from upland areas is concentrated in the root zone. Nevertheless, the plant is still xeromorphic and xeroanatomic, shaped by the action of the extreme climatic factors, *nota bene*, to a lesser degree.

Another category of woody species in arid regions avoids drought during the rainless hot season by dropping their foliage and falling into an absolute or partial dormancy (*Anagyris foetida*).

From the march of the life phenomena of woody evergreen drought enduring xeromorphic species, in the case where rains are the only source of water, the following are their characteristic properties:

1. After the available soil moisture is consumed during the first part of the growing season, and an absolute or a partial equilibrium is obtained, the woody plant continues to maintain its green foliage due to its ability to absorb atmospheric water. Full turgor is maintained in the foliage and in the root system, and a reasonable hydration is found in the xylem and in other organs of the tree. Against the superfluous sources of energy, the xerophyte is isolated by its xeromorphic structure, and by the biochemical properties of the plasma, such as decreased dispersion, increased permeability, increased viscosity and eleasticity, denaturation of the protein and coagulation, and an increase in its osmotic pressure, by a rise in the concentration of sugar.

In damp climates with sufficient water in the soil, stable root systems are formed with regard to depth and ramification. In a mixed forest, the root systems of the various trees develop according to species. In a dry climate, however, such stability is not encountered, but there is marked hydrotropism—the woody plants may develop superficial roots or send out roots which penetrate deeply into the soil where water is available and yet, they are capable of suddenly developing superficial roots after many years of such growth, if agricultural crops are irrigated in the vicinity.

2. The root systems in such conditions are characterized by their superficial and horizontal dispersal, starting from a depth of 5–10 cm. A part of the roots penetrates to the depth to which rainwater penetrates. Below it an eternal dryness rules: therefore, the moisture percentage beneath such a depth is identical under the canopy and in neighboring bare areas.

If available soil moisture is found in deeper layers, xerophytes penetrate with their roots often very deep, as for instance, *Alhagi camelorum*, common in Asian desert, the roots of which reach a depth of 25 meters, and *Glycyrrthiza glabra* 10–15 meters, *Andina* in Brazil 18–19 meters, and *Prosopis* in Arizona desert 19–20 meters (DAUBENMIRE, 1959).

The category of xerophytes which develop a superficial and a horizontal root system elongates its lateral roots to a radius of 40 meters, as *Tamarix aphylla* in the Sahara desert (LADOVER, 1928), and *Larrea mitida* which may extend to a distance of 27 meters.

3. The xeromorphic properties of the xerophytes are accompanied by a specific xero-anatomical structure.
4. In addition to the conventional properties of the xeromorphic leaf, the majority of indigenous desert species and the successfully acclimatized exotics have stomata on both sides of the blade.
5. The most prominent xerophytes can exist for an extended period on dew, mist, and water vapor, but when planted in saturated soils or along rivers they are able to utilize fantastic amounts of water when it is available, as happened when Mediterranean *Tamarix* was introduced to some parts of Arizona. The same phenomenon was stated in regard to the Argentinian *Larrea* which grows naturally in regions where the annual rainfall is less than 100 mm, and flourishes luxuriantly when planted artificially in regions where rainfall is 1000 mm.
6. There is a lack of agreement among workers in regard to transpiration intensity and the degree of xerophytism: some define xerophytes as great savers of water (POOL, 1923); others (MAXIMOV, 1929), measured tremendous transpiration rates, and still others found no great difference in this respect between the two classes of plants (DARWIN, 1914; BLADES, 1935). As far as this problem concerns woody xerophytes grown in an arid environment, transpiration is parallel to other physiological processes and diminishes after the available soil moisture is absorbed by the roots. In spite of the *status quo* in soil moisture, transpiration and the normal phenological phenomena continue until rainfall starts.

DISCUSSION AND CONCLUSIONS

A woody plant within its native habitat or an entirely domesticated one, both grown without the interference of man, are completely adapted and in an absolute harmony with their surrounding environment. The physical environment is the workshop which shapes the morpho-anatomo-genesis of the tree and its form of life. A plant growing in its natural habitat without the interference of man is completely adapted to the surrounding environment after an evolution of thousands and tens of thousands of years. In such cases, one does not encounter the well-known phenomena among the cultivated plants. The first are adapted to the very wide amplitude of climatic and edaphic factors. For example, in the case of lack of water or very high temperature in dry regions, the evergreen plant does not wilt temporarily or permanently, and in a cold climate it is not damaged by very low temperatures. The structure of the tree therefore reflects the properties of the environment and vice versa. By knowing the environment, we are able to predict the structure of the plant, its pattern and rate of growth.

Because of this, the leaf, for instance, is not an unfortunate organ exposed to the mercy of climatic factors, but it is adjusted to its ecophysiological function. The structure of the leaf shows an exceptional plasticity, adjusting itself to the changing micro-ecological conditions. It provides excellent isolation to diffusion of water vapor from the stomata when metabolic activity stops, and it regulates stomatal movement in accordance with the intensity of the metabolic status of the tree. A full turgor is always preserved in the foliage, and it never undergoes any conspicuous dehydration which may affect cell turgor even in years of drought.

Under arid environmental conditions, one must distinguish between xeromorphism caused by the extreme climate and that caused by water deficit in the soil or by both factors together. The latter situation is not unusual in subtropical or desert climates. In such conditions, the addition of water to the soil significantly reduces the degree of xeromorphism, especially regarding the density and size of the stomata of the leaf. Examples relating to these conditions can be seen also in sclerophilic Australian forests growing in regions with only 200–500 mm of rainfall during the winter season, and the same xeromorphic characteristics in trees growing in regions with 1000 mm of rainfall well-distributed throughout the year (WOOD, 1934). Obviously, it is to be expected that in the latter case the xeromorphism will be relatively less pronounced.

A relatively high hydration is preserved within the xeromorphic foliage during the driest and hottest months. For this reason no parallelism exists between the rate of transpiration and leaf water deficiency or its moisture percentage during the critical months when extreme atmospheric stress suppresses physiological activity, including transpiration, to their minimal values.

Due to the maintenance of a relatively high level of hydration in the leaves during the most critical months, there are no significant changes in osmotic pressure throughout the year, as shown by BRAUN-BLANQUET and WALTER (1931) in two dominant species in the arid Mediterranean forests—*Pistacia lentiscus* and *Pistacia palaestina*.

The rain water which percolates into the soil in the arid climate during the cool season is mainly absorbed by the roots during the first half of the rainless season when favorable temperatures activate the woody xerophytes. The available moisture is absorbed until June or the beginning of July, and then a partial or absolute *status quo* is maintained in the root zone, usually higher than in the neighboring bare areas. The amount of surplus water in the root zone of different evergreen trees during the critical months differs for the same species according to soil depth. Woody plants on stony slopes indicate a greater surplus compared to that in bare soil, and compared to the same species growing on deep soil with a relatively greater supply of moisture. The evergreen xerophytes maintain their green foliage and a full cell turgor in the leaves, trunk, and roots is preserved for 3–5 months until rainfall starts. This is possible because of their ability to absorb atmospheric moisture if other sources of moisture are not available in the soil.

The xeromorphism of this type of plant is much more extreme than in the case of the same plant growing nearby in identical microclimatic conditions but in soil with a supply of available water throughout the growing season.

The accepted methods of calculating the quantity of dew and mist absorbed by the plant based on temperature, leaf water potential, and vapor pressure gradient from air to leaf, do not reflect the conspicuous quantities of water utilized by the tree during 4–5 months after a *status quo* in soil moisture is established. They do not reflect the behavior of the desert xerophytes and other evergreen species of the semi-arid regions, which continue to grow in years of drought, when the scarce quantity of rainwater penetrates only to a depth of 40–50 cm, while the roots are 2–2.5 m deep. Moreover, the *status quo* in the root zone is not less during years of drought than during years of normal rainfall.

As shown by GINDEL (1966), the studied tree xerophytes absorb from dew and mist nearly 4/5 of the water necessary for transpiration during the 3–5 critical months (June–October, or July–September, dependent upon yearly rainfall and its distribution) after a *status quo* in soil moisture rules within the root zone.

According to SLATYER (1964), the stomata are generally closed at night. Therefore, the amount of dew absorbed by the plant should suffice only to resaturate the leaf tissues (particularly when the high resistance of the cuticle in dry climates is taken into consideration). But this appears to be invalid in arid conditions. As shown in this investigation, the behavior of stomata at night in 78 tree species was such that the opening of the guard cells started as soon as dew appeared and often before its moisture is felt on the leaf surface at 8:00 or 9:00 p.m. The openings do not increase for many hours, at least until 3:00 a.m., when night experimentation on stomata has been usually interrupted by us. The opinion of SLATYER (1967), that a part of condensed water vapor comes up from the soil, and therefore is not any net gain of water to the plant water balance, is negated by the following facts:
1. If the evaporated soil moisture were not converted into dew, the evergreen xerophytes could not survive during the 3–5 critical months, and during years of

drought in the desert, often rainless all the year round. 2. The evaporation of soil moisture during the night appears in open areas beyond the root zone and therefore it is a net gain to the water balance, due to which the evergreen woody vegetation survives in a dry climatic and edaphic environment.

It appears that soil water deficiency and arid climatic conditions may result in morphological modifications, among which is an increase of stomatal density per unit area. This phenomenon is not necessarily always accompanied by a simultaneous decrease in the leaf area.

The participation of atmospheric water in the control of water shortage in the woody xerophytes and in the reduction of moisture tension in the soil was followed in three ways:

1. By comparison of soil moisture in forest plantations with that in neighboring bare ideas of identical make-up.
2. By comparing the fluctuations in soil moisture in the two areas during the dry and rainy seasons.
3. By investigation of the process of entry of water into the leaves, its passage into the stem and roots, and the egress of the escess into the soil surrounding the roots.

ARVIDSSON's opinion (1951), that the excess moisture under *Tamarix* derives from the dripping of dew from the foliage, is groundless since the 0–30 cm layer is the driest layer. Only a few mm of soil are wetted even after a night of heavy mist, and this moisture disappears during the first few hours of sunshine.

The possibility that the excess moisture in the forest floor is a result of lessened solar illumination and wind, which reduce loss by evaporation in the shade is imaginary for the following reasons: the significant moisture excess to a depth of 210 cm is also present in successive dry years when the scarce quantity of rains (32 mm) managed to wet the upper 30–40 cm of the soil; it is found beneath every endemic tree, shrub, or semi-shrub, growing in isolation and completely exposed to sunlight and winds, provided the plant is green and does not drop its leafage during the critical period. The forest canopy, of course, does provide protection against direct illumination and wind, but even so, the upper 0–30 cm of the soil in a tamarisk plantation are as dry as in the bare area and soil moisture excess reaches its maximum at a depth of 90–120 cm. It follows that the excess moisture in the soil and its distribution are caused by the physiological activity of the trees and not by shade.

Chemical analysis has shown that the water from mist and dew is not pure, distilled water as had been thought. It contains ions that are important in the nutrition of the plant. The amount of chlorine is small and the pH is low (Table X). The source of these ions is the dust particles in the air, resulting from evaporation from the seas and oceans around Israel and from the deserts adjoining the more humid areas of the country.

The quantity of ions present in the rain was measured in Australia by HUTTON (1958), who established that it is greatest in the summer. It proves that, in the summer, the raised dust particles are so tiny (5–10 microns in diameter) that they are unaffected by gravity, and form nuclei for the condensation of raindrops. A similar phenomenon may occur during mists or nights of "heavy dew" when dew forms in the air a fog layer near the ground. MENCHIKOWSKY (1924), in Israel, showed that the amount of chlorine in the rain increases with proximity to the Mediterranean Sea.

Further research is necessary to prove, unequivocally, that the positive and negative ions are expelled by the roots and that their entry into the soil is not only due to simple percolation.

The stomatal constellation and structure are specific to the extreme climatic conditions. The most striking conclusion from the investigation on stomata is their presence on both sides of the leaf in the majority of the indigenous woody species grown in the desert in dry and often rocky soil.

If the problem of stomata is to be considered according to the endemic vegetation, the overall conclusion is that if the climatic and soil drought conditions become extreme, there is an increase in their number.

In the endemic evergreen sclerophytes grown in the sub-tropical Mediterranean woods, as for instance, *Quercus calliprinos*, *Phillyrea media* and others in which the stomata are generally found only on the lower epidermis, the most active leaves turn their lower epidermis upwards at night, and during the growing season they remain in that position, exposed to radiation, throughout the day.

Species having stomata on both sides of the blade are xerophytic and grow naturally under extreme ecological conditions within the desert where the water table is far from the root zone, such as *Balanites aegyptiaca*, *Acacia tortillis* and *A. spirocarpa*. Two of the three local Pistacia species which have stomata only on the lower epidermis (*P. lentiscus* and *P. palaestina*) grow in the relatively more moderate Mediterranean maqui, whereas *P. atlantica*, which has somata on both sides of the blade, grows naturally also in the semi-desert, often on rocky slopes.

In the most successfully acclimatized exotic tree species morphological and anatomical modifications were noted, as a result of which 300 species adapted to conditions more arid than those of their original habitat (GINDEL, 1957). Such modifications were more clearly evident the greater the differences between the climatic and edaphic conditions of their native habitat and those of Israel. In the majority of the exotics investigated and successfully acclimatized, stomata are present on the upper and lower epidermis—particularly the Myrtacae (*Eucalyptus sp.*, *Melaleuca sp.*, and *Calistemon sp.*) and *Acacias*.

This fact is true not only for the endemic and acclimatized vegetation which grows in Israel's desert, but also for the desert vegetation in the United States, central Australia, Africa, and Asia. Most of the plant species which predominate in these places and which grow under very dry conditions, have stomata on both sides of the leaf.

According to WOOD (1934), 100% of the plant species in the arid region of Koonamore, and 88% of the sclerophilic forests in Victoria have stomata on both leaf surfaces. Of 39 sclerophilic species growing in south Australia which WOOD examined, 30 had stomata on both leaf surfaces, and of these, three species had more stomata on the upper epidermis, and in another two, the stomata were found only on the upper epidermis (*Cheirvanthera linearis*, *Billardiera scandens*).

In 25 species, the number of stomata on both sides of the leaf was the same. In another location in southern Australia, the stomata of desert shrubs were found on both sides of the leaf of 28 species, and in three species there were more stomata on the upper epidermis (*Eremophila sturtil, Myoporum platycarpum, Eucalyptus longifolia*). The generally accepted theory that tropical plants and those found in humid regions have a greater number of stomata than plants from arid zones is incorrect. On the contrary, the opposite is true. Our studies with woody plants in the forests of central Europe showed stomata only on the lower epidermis. A similar situation was found by SALISBURY (1928), in woody plants of England's forests.

Moreover, in some species the leaves developed during the hottest months (July, August) have a denser net of stomata on the upper epidermis than leaves developed during the months of February–May (GINDEL, 1968). These facts do not coincide with expectations from the ecophysiological point of view. Such a characteristic constellation of stomata does not act to increase transpiration but to increase the ability of the leaf to absorb atmospheric moisture.

These results may satisfy STOCKER's (1960) anxiety which was revealed when he concluded his survey on stomata saying: "The increase of frequency of stomata and veins in a situation in which water must be saved, will, in this respect, remain problematic."

The stomata constellation is not a simple pipe complex of physical properties, but a very complicated biological system. Stomata behavior differs, even between neighboring stomata during daylight and night, in regard to the size of the openings and their form, or between the stomata of the upper and lower epidermis during the day and night. There are dynamic and heterogeneous changes in the position of the guard cells in response to solar radiation, to fluctuations in available soil moisture and temperature beginning at sunrise and lasting until sunset. The various orientations of neighboring leaves in the same plant (particularly in the genus Eucalyptus) reveal many problems which are still waiting for a scientific clarification.

Even in the case when stomata are closed there is no certainty that they are absolutely closed and not permeable to water molecules whose diameter is only two ångströms. When they are open there is often a decline in physiological activity. WENT (1944), for instance, proved that there is no correlation between the width of the opening and the absorptive power of the leaf, and that the stomata close in the afternoon even when there is sufficient water in the soil. Similarly GREGORY *et al.* (1950), proved that the rate of absorption of CO_2 into the leaf of the tomato was not depressed even the stomata were closed.

In regard to the correlation between the behavior of stomata and transpiration intensity, this study showed that: 1. Xeromorphic leaves of smaller area grown under soil water stress and transpiring sparingly, increase their rate by a factor of 2, and equal or even surpass the rates of well-watered plants, when supplied by artificial irrigation. 2. There is a changing stomata constellation and size within the same plant even during the same growing season together with changes in the intensity of climatic factors. 3. There is a larger area of opening at night than during the day in trees grown in a water-deficient soil during the hottest and driest months.

The succulents, which are characteristic of habitats in which water is usually in extremely short supply, have evolved a biochemical mechanism which enables them to take carbon dioxide in and store it as organic acid during the night when water loss by evaporation is at a minimum. During the day, behind tightly shut stomata, this stored carbon dioxide can then be photosynthesized into sugars without loss of water to the air (FOGG, 1968).

The adjusted stomata constellation, its linkage to the environment, and its dynamic changes abolish the idea of using transpiration suppressants to achieve efficiency of water use in the production of dry matter.

In light of the real definition of the process of transpiration, and of the bio-logical characteristics of the stomata previously described, there is no possibility of reducing the transpiration rate without affecting the rate of the other physio-logical activities. This is so because of the diversity in the behavior of the stomata, even in the case of neighboring stomata, concerning the size and shape of the aperture, the number of open and closed stomata, the relation between density and size to the microclimatic conditions existing at the time of morphogenesis and to the amount of available soil water. Any change in the status of the stomata by partially blocking them disturbs the entire metabolic process. On the other hand, if the purpose is to delay ripening of fruit or other yeilds of agricultural plants, or to retard growth of forest trees, then transpiration suppressants may fulfill an important role.

The definitions of transpiration by different scientists over a period of 245 years, beginning with the published results of HALES (1727), do not reflect realistically the biological characteristics in general and the physiological ones in particular.

The physical-meteorological approach to plant-water relationships, which simplifies this problem by identifying it with physical evaporation, did not advance the science concerning the process of water consumption by plants. It will be useful to site the conclusions of LEMON's work (1967), published almost 40 years after the appearance of Thornthwaite's theory on potential evapo-transpiration: "The more we learn about the physics of energy exchange and physiology of plants, the less we can say about water use efficiency in quantitative terms. . . . The extent to which we can use that knowledge to improve the water use efficiency in crops remains a challenge."

The physical laws defining the rate of evaporation as a function of solar energy, temperature, and wind velocity, are not identical with the biological laws defining the rate of transpiration.

Since transpiration is a physiological process, it is not bound by physical laws related to physical evaporation. This fact is particularly outstanding in woody plants, xerophytic in nature, which exist independently. We do not witness any signs of wilting, and thus the contention that transpiration is able to desiccate the leaf tissues (KOZLOWSKI, 1964) is contradicted.

MAXIMOV's opinion (1929) does not agree with the situation which we found under arid conditions for woody xerophytes. His contention was that drought resistance is reflected in a capacity to endure permanent wilting. It follows that evergreen xerophytic woody plant growing in arid habitats excel in their ability to retain full turgor by reducing the water tension in the cells by absorbing atmospheric moisture. We have never found signs of wilting anywhere in the desert,

126

even in drought years with only one fourth of the annual rainfall. For this reason we also did not find any conspicuous dehydration of the xerophytes we examined. This also explains why even the most prominent xerophytes which do not go into dormancy by shedding leaves and stopping their physiological activity, and which retain their green foliage at full turgor all the year round, are unable to exist without water. If moisture is unavailable in the soil, it is absorbed from the air.

These results show clearly that the idea of cooling the leaf by means of transpiration is invalid because of the reason mentioned before. The xeromorphic and xero-anatomic structure of the leaf reduces physical evaporation by closing the stomata and eliminating cuticular transpiration (STOCKER, 1960); in addition, certain specific properties of the plasma form a suitable barrier to allay the superfluous actions of heat.

THORNTHWAITE and MATHER (1951) claim that transpiration utilizes excess heat energy and prevents the plant from becoming overheated.

The lack of a realistic basis to the contention that the role of transpiration is to cool the leaf and prevent overheating is indicated not only by the results obtained from studies in the arid zones of Israel, but also in the arid part of Australia (HELLMUTH and GRIEVE, 1969 and HELLMUTH, 1970), and in the Death Valley of the United States (STARK, 1969). These results show that during those months in which the temperature, the radiation, and the hot dry winds reach maximum values, the transpiration rate drops to a minimum while at the same time an equable hydration level is maintained in the leaves. HELLMUTH even emphasizes that plants most tolerant to high temperature grow in the most arid zones.

Only a biological approach to the problem will lead us to a rational solution. There is an optimum constellation of climatic and edaphic factors which is a function of the biological properties of the species. This optimum constellation shapes the morpho-anatomo-physiological complex and metabolic activity of the plant.

These results are in disagreement with the opinion of CURTIS (1926), and THOMAS (1954), that the greatest part of the water absorbed from the soil is lost by transpiration without participating in the metabolic activities of the plant. On the contrary, the transpiration rates in this instance reflect the intensity of metabolic activities.

The opinion that only a small percentage of water (1–3%) is used in photosynthesis activity, and that the rest absorbed by the roots ("the inert pumps") is vaporized in response to atmospheric energy, does not coincide with the behavior of the xerophytes in the extreme climatic conditions in which this investigation was conducted. In order for photosynthesis to take place, 80–90% (on a fresh weight basis) of the photosynthetic tissues must be water (FOGG, 1968).

Water equilibria serve as important integrating forces in a plant life (WOODHAMS and KOZLOWSKI, 1954; KOZLOWSKI, 1964, 1968; KRAMER, 1949, 1963). When the direct and indirect participation of oxygen and hydrogen in plant life and in its structure are considered, it follows that each molecule of water entering the plant fulfills this or that function, directly or indirectly, in the metabolism and that the water vapor transpired is its product. Adequate hydration in the cells of the plant and within the protoplasm is necessary for all the life processes: a water deficiency affects all of them simultaneously. Without a proper hydration of the protoplasm no translocation of raw materials from the soil to the foliage, and no movement of photosynthesis products, can take place.

The body of the trunk is mainly composed of molecules of water. The number of molecules of water used to construct the xylem of the trunk indicates the extent to which they contribute to the energy of growth in the tree's lifetime. The xylem is generally composed of 45% cellulose, 25–30% hemicellulose, 2–5% other carbohydrates (pectin and starch), and 25% lignin. The water molecules comprise 56% of the cellulose, 55% of the hemicellulose, 54–56% of the sugars, and 35% of the lignin. Thus, an average of 51% of the xylem's weight consists of water. Every product of carbohydrate metabolism carries hydrogen and oxygen, and most of the constituents of cell walls and protoplasm have these elements in their makeup (CRAFTS, 1968).

Water is necessary to protein synthesis and enzymatic activity, mineral relations, nitrogen, metabolism flow of oleoresin and latex, and other processes (KRAMER, 1949; ARMY and KOZLOWSKI, 1951; KOZLOWSKI, 1964).

In addition to the relation between transpiration and the different physiological activities which have been outlined in this book, one must not forget the relation between ion uptake and transpiration. With an increase in transpiration ions are carried along by mass flow in the transpiration stream (HOAGLAND, 1944; BRAYER, 1951; LUNDEGARTH, 1950; BROUWER, 1954; HYLMO, 1957; CANNING and KRAMER, 1958; PETTERSON, 1960; KOZLOWSKI, 1964).

The metabolic water exhaled through the stomata and lenticles of the plant reflects the rate of physiological activity and therefore its rate is parallel to photosynthesis, cambium activity, growth, enzymatic activity, the formation of amino acids and fats, and to other life phenomena as detailed in this investigation. The amplitudes of water, temperature, and light, within which transpiration reaches its optimal values, coincide with the amplitudes of other physiological processes. In fact, since transpiration is a function of plant activity, it reflects the dynamic nature of life; this is why it is as variable as other physiological processes. Prolonged transpiration signifies prolonged, active metabolism, leading to strong growth.

The accepted definition of the plants requirements of water concerns only the vapor exhaled through the stomata, but the moisture entering them is ignored. When physiological activity takes place there is a constant entry of oxygen and carbon dioxide accompanied by water vapor, so that metabolism is accompanied by a two-way gaseous exchange.

There is a fundamental biological observation well known for many years that the source of the energy absorbed by chlorophyll is a small fraction of the energy of sunlight (ALLEN, 1963). In addition, as mentioned previously, the quantity of solar energy used for transpiration, which is a link of metabolic activity, is dependent upon several biological properties, e.g.:

1. The trees can be light-demanders, very tolerant species, or even shade-demanders.
2. The plants are physiologically active or inactive.

Already in 1905 LUBIMENKO proved that shade plants are capable of producing the same amount of organic matter with relatively less light than light-demanders.

In shade plants in temperate climates photosynthesis begins in light of 1080 lux, while light-demanders require 4300–5400 lux. Some species show greater differences in reflection for light in the wavelengths beyond 0.7 micron than in the visible range.

Direct solar radiation is not always necessary for tolerant species and shade-demanders. They often show a high growth rate even in the presence of diffuse light (*Fagus sylvatica* or *Picea excelsa* on the northern slopes of the Alps). The giant redwood has a very low light requirement while the Engelmann spruce and Douglas fir approximately twice as much light is necessary for appreciable growth (BATES and ROESTER, 1928). Sequoia seedlings are known to assimilate carbon dioxide in light as weak as 1080 lux on the floor of the forest which is completely cut off from direct sunlight by the dense cover of foliage.

According to TRANQUILLINI (1954) shade leaves of beech produce 4–5 times as much carbohydrate as sun leaves. BOURDEAU and LAVERICK (1958) found that shade leaves of white and red pine have higher rates of photosynthesis per unit of leaf weight than sun leaves.

3. In arid climates some additional biological factors interfere with the mentioned physico-meteorological approach, namely the highly specialized evergreen xeromorphic structure of the foliage and the absorption of atmospheric moisture at darkness.

The studies of plant–water relations led to unorthodox results and indicated the need to change our approach concerning a number of basic biological characteristics from the morphological, anatomical, and ecophysiological points of view.

The function of solar radiation in plant life is not to drain out the water from the foliage and to exploit its energy. On the contrary, all the energy of life originates from the sun by the process of photosynthesis during which carbohydrates are manufactured. These organic compounds serve as the primary energy sources for the life phenomena of the plants. The water vapor and gases released during photosynthesis by the process of transpiration are the products of metabolism.

BLACKMAN in 1905 (PANARES, 1967) asserted that the rate of photosynthesis increases as the amount of light increases, and it levels off at a peak value beyond which higher light intensity has no effect on the rate of photosynthesis. That peak is, of course, dependent on species, whether it is a light-demander or shade-bearer, as well as on the quantity of available water and soil fertility. The lower the fertility and available water, the higher is the demand for solar light for the process of growth.

The importance of the interaction of biological properties with the physical environment in limiting metabolism, hence transpiration, is illustrated by these tree xerophytes in which biological controls override the intensity of the environmental factors.

Solar energy, as other ecological factors, is a link in the assembly of the climatic and edaphic factors and dependent upon biological properties of the species.

The radiation energy may be as extreme as possible, but the transpiration will nevertheless become slightest when the physiological activity drops to its lowest values. Moreover, the more fertile the soil, the lower the optimal quantity of light necessary to attain a maximum transpiration. For example, in fertile forests, transpiration reaches 35″ during the seasons of growth, whereas in poor forests it is 5–15″ (MINCKLER, 1939). The annual transpiration rate may vary from less than 1″ where pioneer species grow on the worst sites to more than 30″ in well stocked climax forests of the best site quality (KITTREDGE, 1948). Besides, the xerophyte are able to preserve a minimum of moisture within the root zone that is always higher than that in the bare area irrespective of the intensity of solar energy and the yearly rainfall. Table XXVIII shows the soil moisture percentage measured at

Table XXVIII. Mean Soil Moisture (in % by weight) within the Forest and in Bare Area at the End of the Dry Season. *Tamarix aphylla*, Revivim.

Date of testing	18.10.60	21.10.61	13.10.63	1.11.64
Rainfall in mm and number of rainy days	44.1(16)	104.4(32)	32.8(8)	164.9(33)
Soil moisture in %: within the forest	2.79	2.36	2.25	1.80
In the open area	1.02	1.09	0.73	0.98

the end of the dry season during 4 years in a *Tamarix aphylla* planting, 23 years old, with 580 trees per ha, and in an identical bare area in the neighborhood (GINDEL, 1967). The planting is located in the desert, in sand, where the yearly rainfall is 100 mm. The same solar radiation in the forest and the bare area caused an evaporation almost twice as high in the latter place.

According to the opinion of THORNTHWAITE and KENNET (1932), the external energy is supplied to the evaporating surface principally by solar radiation. Some of the radiation is reflected back from the surface, the percentage lost being known as the albedo.

The question which arises is: does the amount of albedo and the reflected energy returned to space depend on the physical properties of the vegetation type or on ecophysiological conditions?

According to the opinion of STANHILL (1961) the radiation balance is the most important climatic factor determining the amount of evapotranspiration. However, it follows from our investigation that this statement is true in regard to physical evaporation, but not as far as it concerns the process of transpiration. STANHILL (1961) made an investigation on Mt. Carmel in Israel. He used a helicopter, flying 10 meters above the tops of four vegetation covers during four different seasons, and measured the net radiation flux to the area which presents the amount of available energy. All the values are mean reflectivity percentages of the appropriate incident solar radiation. His values (Table XXIX) suggest that the amount of energy available for evapotranspiration, or in other words, for water loss by evaporation, is greatest for the pine forest, next the maqui scrub, followed by natural pasture, and least in the case of the non-irrigated winter grain.

The measured rates of transpiration and intensity of physiological activity carried out in the same area (GINDEL, 1964, 1967a, 1968, 1971) do not coincide with the net radiation values. The differences in the various types of vegetation measured by STANHILL reflect rather the physical properties of the plant cover, such as color, density, etc.

The direct radiation falling on the leaf surface is reflected, absorbed, or transmitted. From this point of view, the question arises to what extent the woody plant is able to adapt itself to the radiation in order to survive under extreme climatic conditions and dry soil prevailing in deserts or semi-arid zones. This matter is particularly unclear regarding xerophytic trees. The work by HOWARD

Table XXIX. Mean Percentages of the Reflection of Incident Solar Radiation by four Vegetation Types (STANHILL, 1961)

Vegetation types	Date of measurement			
	January 16	March 23	July 21	September 28
Pine forest	0.18	0.20	0.17	0.18
Maqui scrub	0.24	0.21	0.20	0.22
Natural pasture	0.29	—	0.22	0.22
Non-irrigated winter grain	0.26	0.26	0.26	0.28

(1966) with different species of Eucalyptus has thrown some light on the subject. He studied the effect of the visible and near infrared spectrum (4,000–9,500 Å), and concluded that reflection by these different sclerophytic species is higher and enables the plant to withstand extreme climatic conditions.

There is no doubt that the leaf's orientation to the incident solar flux is an important means for the plant to overcome extreme radiation conditions, and by adjustment of the proper leaf angle to the radiation, the transmission intensity is reduced. If the angle in relation to the incident radiation increases, the transmission is reduced and the reflection increases (SHOLGIN, 1960). PENMAN (1966) arrived at a similar conclusion, and showed that in the case of endemic species in western Australia the reflectance values grow when the water percentage drops from 68.4 to 50.3%. BILLINGS and MORRIS (1951) found that some desert plants reflect strongly in the infrared. In addition to all these, it is also known that transmitted light is deficient in photosynthetic light (FOGG, 1968).

In order to know if radiation indeed causes the development of a complicated pattern of angles in the foliage of different Eucalyptus trees, we conducted an experiment with a number of 20-month old species growing in pots, of *Eucalyptus camaldulensis, E. gomphocephala* and *E. forrestiana*. One group of each species was held in a shade house with homogeneous light, and a second group in full light. We found that in the first group almost all the leaves of the 100 plants of each species were oriented horizontally, while on those plants in full light the leaves were turned in many different directions in relation to the radiation.

BIBLIOGRAPHY

ACKLEY, W. B. 1954. Seasonal and diurnal changes in the water deficits of Bartlett pear leaves. *Plant Physiol.* 29:445–448.

ALLEN, J. M. 1963. The nature of biological diversity. McGraw-Hill Book Company, p. 304.

ANGUS, D. E. and BIELORAI, J. 1965. Transpiration reduction by surface films. *Aust. J. Agric. Res.* 16:107–112.

ARMY, T. Y. and KOZLOWSKI, T. T. 1951. *Plant Physiol.* 26:353–365.

ARVIDSSON, J. 1951. Austrocknungs- und Dürreresistenzverhältnisse einiger repräsentanten öländischer Pflanzenvereine nebst. Bemerkungen über Wasserabsorption durch oberirdische Organe. *Oikos. Acta oecologica scandinavica* Suppl. 1, p. 181.

ASHBEL, D. 1936. On the importance of dew in Palestine. *J. Palestine Oriental Soc.* 16:316–321.

ASHBEL, D. 1949. *Geog. Review,* 39:291–295.

BARR, C. J. 1949. Photosynthesis in maize as influenced by transpiration spray. *Plant Physiol.* 10:86–97.

BATES, C. J. and ROESER, J. 1938. Light intensities required for growth of conifer seedlings. *Am. J. Botany* 15:184–195.

BAUMGARTEN, A. 1934. Thermoelektrische Untersuchungen über die Geschwindigkeit des Transpirationstromes. *Z. Bot.* 28:81–136.

BILLINGS, W. D. and MORRIS, R. J. 1951. Reflections of visible and infrared radiation from leaves of different ecological groups. *Am. J. Bot.* 38:327–331.

BILLINGS, W. D. 1952. The environmental complex in relation to plant growth and distribution. *Quart. Rev. Biology,* 27:251–265.

BLAYDES, G. W. 1935. Water vapour loss from plants growing in various habitats. *Ohio J. Sc.* 35:112–130.

BONNER, J. and GALSTON, A. W. 1952. Principles of Plant Physiology, Freeman, San Francisco, p. 630.

BOURDEAU, P. F. and LAVERICK, M. G. 1958. Tolerance and photosynthetic adaptability to light intensity in white pine, red pine, hemlock and Ailanthus seedlings. *Forest Sci.* 4:196–207.

BRAUN-BLANQUET, J. and WALTER, H. 1931. Zur Ökologie der Mediterranpflanzen, *J. B. Wiss. Bot.* 74:697–748.

BRAWDO, B. A. 1972. Effect of several transpiration suppressants on carbon dioxide and water vapor exchange in Citrus and grapevine leaves. *Physiologia Plantarum,* 26:152–156.

BREZEALE, S. J. *et al.* 1950. Moisture absorption by plants from an atmosphere of high humidity. *Plant Physiol.* 25:419–423.

BRIX, M. 1962. The effect of water stress on the rate of photosynthesis and respiration in tomato plant and Loblolly pine seedlings. *Physiol. Plantarum* 15:1–20.

BROUWER, R. 1953. *Acta Bot. Neerl.* 3:264.

BROYER, T. C. 1951. *Plant Physiol.* 26:655.

BRAUN, V. A. 1963. Photosynthesis and transpiration upper and lower surfaces of intact banana leaves. *Plant Physiol.* 36:399–404.

CAVERA, A. L. 1955. Latin America. Plant Ecology, UNESCO, p. 376.

CIAMPI, C. 1954. Ritmo di accrescimento del legno e caratteristiche anatomiche in individui di abette bianco (Abies alba Mill.) di diversa provenienza. L'Italia Forest. a Monte, 6.

CLEMENTS, H. F. 1934. Significance of transpiration. *Plant Physiol.* 9:165–172.

CRAFTS, A. S., CURRIER, H. B. and STOCKING, C. R. 1949. Water in the physiology of plants. *Chronica Bot. Comp.* p. 240.

CRAFTS, A. S. 1968. Water structure on and water in the plant body. pp. 23–47. Water Deficit and Plant Growth, by J. J. Kozlowski. Academic Press, p. 333.

CRAGG, J. B. 1968. Toward understanding Ecosystems. *Advances in Ecological Research* 5:1–35.

CURTIS, O. F. 1926. What is the significance of transpiration? *Sci.* 65:267–271.

DARWIN, F. 1898. Observations of stomata. *Phil. Trans. R. Soc. B.* 190:531.

DARWIN, F. 1914. The effect of light on the transpiration of leaves. *Proc. Roy. Soc. B.* 89:281–299.

133

Daubenmire, R. F. 1959. Plants and environment. p. 422.

De Witt, C. T. 1958. Transpiration and crop yields. *Jns. of Biol. & Chemical Research*, Wageningen. Lanbouwk. onder L. No. 64. 6, The Hague.

Dvoretskaya, E. J. 1949. On the drought-resistance of the summer oak and certain other genera. Forest Economy, No. 12.

Duvdeuvani, Sh. 1947. *Quart. J. Roy. Meterol. Soc.* 73:282.

Eckhardt, F. E. 1952. Rapports entre la grandeur des feuilles et le comportement physiologique chez les xérophytes. *Physiologia Plantarum* 5:52–69.

Eckhardt, F. E. 1960. Eco-physiological measuring techniques applied to research on water relations of plants in arid and semi-arid regions. UNESCO; *Arid Zone Res.* 15:139–171.

Eisenzopf, R. 1952. Ionenwirkungen auf die cuticulare Wasseraufnahme von Conifers. *Phytom* 4:149.

Evenari, M. and Richter, R. 1937. Physiological and ecological investigations in the Wilderness of Judea. *T. Linnean So. Botany* 51:333–381.

Fahn, A. 1964. Some anatomical adaptations of desert plants. *Phytomorphology* 14:93–102.

F.A.O. 1962. Forest influences. p. 307.

Fog, G. E. 1968. Photosynthesis. American Elsevier Publishing Company, Inc., New York. p. 116.

Gardner, W. R. 1965. Soil, water movement and root absorption. Plant Environment and efficient water use. *Amer. Soc. Agronomy*, pp. 127–149.

Gates, D. M. 1963. Leaf temperature and energy exchange. *Arch. Met. Geophys. Bioklim.* 12:321–336.

Gates, C. T. 1968. Water deficits and growth of herbaceous plants. Water deficits and plant growth. (T. T. Kozlowski.) Academic Press, p. 333.

Gilead, M. and Roseman, N. 1954.

Gindel, I. 1944. Aleppo pine as medium for tree-rings analysis. *Tree-ring Bulletin*, Arizona. 11:6–8.

Gindel, I. 1947. Ricerche sui semi di specie forestali indigene ed introdotte in Palestina. Ist. di Selvicoltura del l'Univ. di Firenze. p. 48.

Gindel, I. 1951. The Eucalyptus in Israel. Results of experiments in acclimatization of Eucalyptus during the years 1931–51. *Agr. Res. Sts. Bull.* 43:68.

Gindel, I. 1952. Some anatomical features of the indigenous woody vegetation in Israel. *Bull. Res. Council of Israel* No. 2, pp. 1–16.

Gindel, I. 1956. Acclimatization of woody plants. "Am Oved". Tel-Aviv. p. 355.

Gindel, I. 1957. Acclimatization of exotic woody plants in Israel: the theory of phytoplasticity. *Materiae Vegetabilis*, Holland, 2:81–101.

Gindel, I. 1959. A tree-ring analysis of Italian forest trees. *Monti e Boschi.* 10:156–164.

Gindel, I. 1960. Biological function of fruit. *Nature.* No. 4731, 142–147.

Gindel, I. 1960. Ecological behaviour of the coffee plant in semi-arid conditions. *Materiae Vegetabilis*, 2:81–101.

Gindel, I. 1961. The water in xerophytes. Proc. XIII Congr. I.U.F.R.O. 1:2–6.

Gindel, I. 1964a. Seasonal fluctuations in soil moisture under the canopy of xerophytes and in open areas. *Commonwealth Forestry Review.* 43:219–234.

Gindel, I. 1946b. Transpiration of Pinus halepensis as a function of environment. *Ecology* 45:868–873.

Gindel, I. 1966. Attraction of atmospheric water by woody xerophytes in desert and semi-desert conditions. *Empire Forestry Review* 48:217–242.

Gindel, I. 1967. The relation between transpiration and six ecological factors. *Oecologia Plantarum* 2:227–239.

Gindel, I. 1968a. Some ecophysiological properties of tree xerophytes grown in desert. *Oecologin Plantarum* 2:49–67.

Gindel, I. 1968b. Dynamic modifications in alfalfa leaves growing in sub-tropical conditions. *Phys. Plantarum* 21:12–87.

Gindel, I. 1969a. Stomatal number and size as related to soil moisture in tree xerophytes in Israel. *Ecology* 50:263–267.

Gindel, I. 1969b. Making the desert bloom. *American Forests* 75:32–37.

Gindel, I. 1970. The nocturnal behaviour of stomata of xerophytes grown under arid conditions. *The New Phytologist.* 69:2.

Gindel, I. 1971. Transpiration in three Eucalyptus species as a function of solar energy, soil moisture and leaf area. *Physiologia Plantarum* 24:143–149.

Gregory, F. G. and Miltrope, F. L., Pearce, H. L. and Spencer, H. Y. 1950. Experimental studies of the factors controlling transpiration. *Journ. Exptl. Bot.* 1(1):1.

GRIEVE, B. Y. and HELLMUTH, E. O. 1970. Ecophysiology of Western Australian plants. *Oecologia Plantarum* 5:33–68.

HALES, S. 1738. Vegetable Staticks. Vol. I 3rd ed. 375 p.

HALEVY, A. 1956. Orange leaf transpiration under orchard conditions. Influence of leaf age and changing exposure to light on transpiration, on normal and dry summer days. *Bull. Res. Council of Israel S.D.* (2–3):154–181.

HALEVY, A. 1959. Diurnal fluctuations in water balance factors of gladiolus leaves. *Bul. Res. Counc. Israel*, 80:239–246.

HANKS, R. J., GARDNER, H. R. and FLORIAN, R. L. 1967. Plant growth-evapotranspiration relations for several crops in the Central Great Plains. *Agronomy J.* 61:30–34.

HEINICKE, A. I. and CHILDERS, W. F. 1936. The influence of water deficiency in photosynthesis and transpiration of apple leaves. *Proc. Amer. Soc. Hort. Sci.* 33:155–159.

HELLMUTH, E. O. 1968. Ecophysiological studies on plants in arid and semi-arid regions in Western Australia. *J. Ecol.* 56:319–344.

HOAGLAND, D. R. 1944. Lectures on the Inorganic nutrition of plants. *Chronica Botanica.* Waltham, Mass.

HOOVER, M. D. 1962. Water action and water movement in the forest. Forest Influences. F.A.O. 15:31–80.

HOWARD, J. A. 1966. Spectral energy relations of isobilateral leaves. *Australian J. Biological Sciences* 19:757–766.

HUBER, B. 1956. Die Transpiration von Sproszachsen und anderen nicht foliosen Organen. Handbuch der Pflanzen-Physiologie 3:427–435.

HUBER, B. 1927. Zur Methodik der Transpiration-Bestimmung am Standort. *Ber. Deutsch. Bot. Ges.* 45:611–618.

HUTTON, I. T. 1958. The chemistry of rainwater with particular reference to conditions in south-eastern Australia. Climatology and Micro-climatology. Proceedings of the Canberra Symposium. UNESCO:285–291.

MYLMO, 1958. *Physiol. Plant.* 11:382.

JAMES, W. O. 1958. Succulent plants. *Endeavour* 17:90–95.

KETELLAPPER, H. I. 1959. *Amer. J. Bot.* 46:225–231.

KETELLAPPER, H. I. 1963. Stomata physiology. *S. Rev. Plant Physiol.* 14:249–270.

KIDD, F. L. and WEST, C. 1920. *Ann. Bot. Lond.* 34:439.

KIENDL, M. 1953. Beiträge zum Wasserhaushalt von Pflanzengesellschaften. *Ac. dtsch. bot. Ges.* 66:248–263.

KITTREDGE, I. 1937. Natural vegetation as a factor in the losses and yields of water. *J. Forestry* 35:1011–1015.

KITTREDGE, I. 1948. Forest Influence. F.A.O. p. 394.

KOZLOWSKI, T. T. 1964. Water metabolism in Plants. Harper and Row, New York.

KOZLOWSKI, T. T. and CLAUSEN, T. Y. 1965. Changes in moisture contents and dry weights of buds and leaves of forest trees. *Bot. Gaz.* 126:20–26.

KOZLOWSKI, T. T. and CLAUSEN, T. Y. 1967. Effects of alkenylsuccinic acids on moisture content of woody plants. *Plant Physiol.* 42, (suppl.), 17.

KRAMER, P. I. 1942. Species differences with respect to water by plants in poorly aerated media. *Am. J. Bot.* 27:216–220.

KRAMER, P. I. 1949. Plant and soil water relationships. McGraw-Hill, New York, p. 347.

KRAMER, P. I. and KOZLOWSKI, T. T. 1960. Physiology of trees, p. 642.

KRAMER, P. I. 1969. Plant and soil water relationships. A modern synthesis. McGraw-Hill, p. 482.

LADAUVAN, L. 1927. Les forêts du Sahara. Rev. des Eaux et Forêts, 6–7.

LANGE, O. L., KOCH, W. and SCHULZE, E. D. 1969. CO_2-Gaswechsel und Wasserhaushalt von Pflanzen in der Negev-Wüste am Ende der Trockenzeit. *Ber. dtsch. Bot. ges. Bd.* 83:39–61.

LARCHER, W. 1960. Transpiration and photosynthesis of detached leaves and shoots of *Quercus pubescens* and *Q. ilex* during desiccation under standard conditions. *Bul. of the Res. Council. of Israel* 8:213–224.

LEE, R. 1968. The hydrological importance of transpiration control by stomata. *Water Resources Research* 3:737–753.

LEMON, E. R. 1963. Energy conversion and water use efficiency in plants. Plant environment and efficient water use. Ed. W. H. PIERRE et al., p. 295.

LESHEM, B. 1968. Adaptation of Pine Roots to an Arid Environment. Doctoral Thesis. Hebrew University of Jerusalem.

LEVIN, M. and REINBENBACH, A. 1956. The chemical composition of Eucalyptus rostrata grown in Israel. *Bull. Res. Counc. Israel*, Vol. 5A, 4:243–252.

LOFFIELD, S. V. G. 1921. The behaviour of stomata. Publ. Carnegie Inst., Wash., p. 314.

135

LUBIMENKO, W. 1905. Sur la sensibilité de l'appareil chlorophyllien des plantes ombrophiles et ombrophobes. *Rev. Gén. Bot.* 17:381–415.

LUNDEGARTH, H. 1950. The translocation of salts and water through wheat roots. *Physiol. Plant.* 3:103.

MACDOUGLAS, D.F. 1910. The water-balance of succulent plants. Carnegie Inst. Wash., Publ. 141:77.

MAKKINK, G. F. 1957. Ekzameno de la formula de Penman, Neth. *J. Agric. Sci.* 5:290–305.

MANSFIELD, T. A. 1965. Studies in stomatal behaviour. *J. Exp. Bot.* 16:721–731.

MAXIMOV, N. A. 1929. The physiological nature of drought resistance of plants. *Proc. Intern. Congr. Plant Sciences* 2:1169–1175.

MAXIMOV, N. A. 1929. The plant in relation to water. Allen and Unwin, London. 451.

MAYER, A. M., POLJAKOFF, A. and MAYBER. 1963. The germination of seeds. International series of monographs on pure and applied biology. Oxford, London, p. 236.

MCGINNIES, W. G. *et al.* 1968. Deserts of the world. Univ. Arizona Press. p. 788.

MEIDNER, H. and MANSFIELD, J. A. 1965. Stomatal responses to illumination. *Biol. Rev.* 40: 483–509.

MENCHIKOWSKY, F. 1924. Composition of rain falling at Tel Aviv. Inst. of Agr. and Natural History Bull. No. 2.

MEYER, B. S. and ANDERSON, D. B. 1956. Plant Physiology. D. Van Nostrand Comp., Inc. p. 984.

MEYER, B. S., ANDERSON, D. B. and BOHNING, R. H. 1960. Introduction to Plant Physiology. p. 541.

MILLER, E. G. 1938. Plant Physiology. McGraw-Hill Book Company, New York and London.

MINCKLER, L. S. 1939. Transpiration of trees and forest. *J. Forestry*, 37:336–339.

MONTEITH, J. L. 1963. Gas exchange in plant communities. In "Environmental Control of Plant Growth". (L. J. EVANS eds.) Academic Press, New York and London, pp. 95–112.

MONTEITH, J. L. 1963. Dew: Facts and Fallacies. The water relation of plants. A symposium of the Brit. Ec. Soc. 1961. p. 394.

NAKAYAMA, M. and KADOTA, M. 1948. The wind influence on the transpiration of some trees. *Bull. Physiograph. Sci. Res. Inst.*, Tokyo Univ. 1:17–34.

NEGISI, K. and SATOO, T. 1954. The effect of drying of soil on apparent photosynthesis, transpiration, carbohydrates reserves, and growth of seedlings of Akamatu (*Pinus densiflora*). *J. Jap. Forest Soc.* 36:66–71.

NEWMAN, E. I. and KRAMER, P. I. 1966. Effects of decenylsuccinic acid on the permeability and growth of bean roots. *Plant Physiol.* 41:606–609.

NISHIDA, K. 1963. Studies in stomatal movement of Grassulacean plants in relation to acid metabolism. *Physiologie Pl.* 16:281–298.

NORTHERN, H. T. 1953. Plant Physiology. The Ronald Press Company, p. 718.

ORSHAN, G. 1954. Surface reduction and its significance as a hydroecological factor. *Journal of Ecology*, 42:442–444.

PANARES, R. R. 1967. Energy, organization and life. Educational Methods, Inc., Chicago. p. 129.

PARKER, I. 1957. The cut-leaf method and estimations of diurnal trends in transpiration from different heights and sides of an oak and a pine. *The Botanical Gazette*, 119:93–101.

PENFOUND, W. T. 1931. Plant anatomy as conditioned by light intensity and soil moisture. *Am. Journ. of Bot.* 18:558–571.

PENMAN, H. L. 1948. The dependence of transpiration on weather and soil conditions. *Soil Sci.*, 1:74–84.

PENMAN, H. L. 1948. Natural evaporation from open water, bare soil and grass. *Proc. Roy. Soc.* A193:120–145.

PENMAN, H. L. and SCHOFIELD, R. K. 1951. Some physical aspects of assimilation and transpiration. *Symp. Soc. Exptl. Biol.* 5:115–129.

PENMAN, H. L. 1956. Evaporation: An introductory survey. *Neth. J. Agric. Sci.* 4:9–29.

PETTERSON, S. 1960. *Physiol. Plant.* 13:133.

PISEK, A. and CARTELLIERI, E. 1932. Zur Kenntnis des Wasserhaushaltes der Pflanzen I. *Sonnenpflanzen. Jahrb. Wiss. Bot.* 75:195–251.

PLAUT, Z., HALEVY, A. H. and SHMUELI, E. 1967. The effect of growth-retarding chemicals on growth and transpiration of bean plants grown under various irrigation Regimes.

POLJAKOFF-MAYBER and MAZER, A. M. 1963. The germination of seeds. Oxford, Pergamon Press, p. 236.

POLSTER, H. 1950. Die physiologischen Grundlagen der Stofferzeugung im Walde. Bayerischer Landwirtschaftsverlag GmbH, Munich.

136

POOL, R. J. 1923. Xerophytism and comparative leaf anatomy in relation to transpiring power. *Bot. Gaz.* 76:221–241.

RABINOWITCH, E. J. 1956. Photosynthesis and Related Processes. Interscience Publishers, Inc., New York, London. 2088.

RUTTER, A. J. 1968. Water consumption by forests. "Water deficits and plant growth". T. T. KOZLOWSKI. Vol. II, p. 333.

RUTTER, A. J. 1968. Evaporation in forests. *Endeavour* 26:29–43.

SALISBURY, E. J. 1928. On the causes and ecological significance of stomatal frequency with special reference to the woodland flora. *Phil. Trans. R. Soc.*, B. 216:1–65.

SALISBURY, F. F. and ROSS, C. 1969. Plant Physiology. Wadsworth Pub. Com., Inc., Belmont, California, p. 747.

SAMPSON, J. 1961. A method of replicating dry and moist surfaces for examination by light microscopy. *Nature* 191:923–933.

SCARTH, G. W. 1929. The influence of H-ion concentration on the turgor and movement of plant cells with special reference to stomatal behaviour. *Proc. Int. Congr. Pl. Sci.* 1:1151–1162.

SCHNEIDER, G. W. and CHILDERS, N. F. 1941. Influence of soil moisture on photosynthesis, respiration and transpiration of apple leaves. *Plant Physiol.* 16:565–583.

SCHOUW, J. F. 1822. Grundträk till en almindeling plantegeografi. Copenhagen.

SCHUBERT, A. 1940. Untersuchungen über den Transpirationsstrom der Nadelhölzer und der Wasserbedarf von Fichte und Lärche. Tharandter Forstl. Jahrb. 90:821–883.

SHIMSHI, D. 1963. Effect of soil moisture and phenylmercuric acetate upon stomatal aperture, transpiration and photosynthesis. *Plant Phys.* 38:713–721.

SHULGIN, J. A., KASANOV, V. S. and KLESHUIN, A. F. 1960. On the reflection of light as related to leaf structure. *Akad. Nauk. S.S.S.R.* 134:471–474.

SLATYER, R. O. 1960. Aspects of the tissue water relationships of an important arid zone species (Ac. aneurea F. MUELL) in comparison with 2 mesophytes. *Bull. Res. Counc. Israel*, 8D:159–168.

SLATYER, R. O. and BIERHUIZEN, I. F. 1964. The influence of several transpiration suppressants on transpiration, photosynthesis, and water use efficiency of cotton leaves. *Aust. J. Ecol. Sci.* 17:131–146.

SLATYER, H. 1967. Plant-water relationships. Academic Press, London and New York. p. 366.

SCHRODTER, H. 1950. Taumenge und Benetzungdauer. *Biol. Zentr.* 69:72–73.

STALFELT, M. G. 1956. Die Stomataltranspiration und die Physiologie der Spaltöffnungen. *Encyc. Plant Physiol.* 3:351–426.

STANHILL, G. 1962. Energy budget measurement by helicopter. *Layaran.* 15:83–89.

STOCKER, O. 1956. Die Abhängigkeit der Transpiration von der Umweltfeuchte. *Plant Physiol.* 3:436–488.

STARK, N. 1967. The transpirometer for measuring the transpiration of desert plants. *Journal of Hydrology* 5:143–157.

STARK, N. and LOVE, L. D. 1969. Water relations of three warm desert species. *Israel J. Bot.* 18:175–190.

STOCKER, O. 1960. Physiological and morphological changes in plants due to water deficiency. *Arid Zone Research.* UNESCO, 15:63–94.

STONE, E. C. 1950. Water absorption from the atmosphere by plant growing in dry soil. *Science* 111:546–548.

STONE, E. C. 1957. Dew as an ecological factor. *Ecology* 38:407–413.

STOUTJESDIJK, Ph. 1970. Some measurements of leaf temperatures of tropical and temperate plants and their interpretation. *Acta Bot. Neerl.* 19:373–384.

SWANSON, R. H. 1965. Seasonal course of transpiration of Lodgepole pine and Engelmann spruce. Intern. Symp. Forest Hydrology. Pennsylvania State Univ. 417–432.

SWORI, E. M. and RAIGONESE, A. E. 1950. Sobre presión osmótica en plantas de las Salinas Grandes. Bol. del. lab. de Botánica, Facultad de Agronomía, La Plata 3:3–4.

THOMAS, M. D. and HILL, G. R. 1937. The continuous measurement of photosynthesis, respiration and transpiration of alfalfa and wheat growing under field conditions. *Plant Phys.* 12:285–307.

THOMAS, M. *et al.* 1954. Plant Physiology, London. p. 692.

THIMANN, V. K. *et al.* 1957. The physiology of forest trees. The Ronald Press Company. New York, p. 678.

THORNTHWAITE, C. W. 1931. *The Geographical Review* 21:633–635.

THORNTHWAITE, C. W. and HARE, K. 1955. Climatic classification in forestry. *Unasylva* 9:51–59.

137

THORNTHWAITE, C. W. and MATHER, J. R. 1955. The water balance. *Climatology* 8:15–23.

TRANGUILLINI, W. Die Lichtabhängkeit der Assimilation von Sonnen- und Schattenblättern einer Buche unter ökologischen Bedingungen. 8th Internat. Bot. Cong. Soc. 13:100–102.

VAN BAVEL, G. H. 1968. Further to the hydrologic importance of transpiration control by stomata. *Water Resources Research* 4:1387–1388.

VAN DEN HONERT, J. H. 1948. Water transport in plants as a catenary process. *Disc. Faraday Soc.* 3:146–153.

VASILEV, T. M. 1931. The water economy of plants in the sandy desert of South-East Kara-Kum. Studies in Applied Botany, Genetics and Selection, Vol. XXV.

WAGGONER, P. E. 1965. Decreasing transpiration and the effect upon growth. Plant Environment and Efficient Water Use. Edited by W. H. PIERRE et al. p. 295.

WAGGONER, P. E. and BRAVDO, B. 1967. Stomata and the hydrologic cycle. National Academy of Sciences. Vol. 57, No. 4. pp. 1096–1102.

WALTER, H. 1951. Einführung in die Phytologie III. Grundlagen der Pflanzenverbreitung. T. E. ULMER, Stuttgart. p. 525.

WALTER, H. 1962. Die Vegetation der Erde in ökologischer Betrachtung. Ver. Gusta Fischer Verlag Jena, p. 538.

WEAVER, R. T. 1942. Water usage of certain native grasses in prairies and pasture. *Ecology* 22:175–192.

WEIHMER, F. Y. 1929. An improved soil sampling tube. *Soil Science* 27:147–152.

WEISEL, Y. 1960. Ecological studies of *Tamarix aphylla* (1). Karst. *The Water Economy Phyton* 15:19–28.

WENT, F. W. 1942. "The dependence of certain plants on shrubs in Southern California deserts". *Bull. Tarrey. Bot. Cl.* 69:100–114.

WENT, F. W. 1944. Plant growth under controlled conditions. *Amer. J. Bot.* 31:597–618.

WENT, F. W. 1956. The role of environment in plant growth. *American Scientist* 44:378–398.

WENT, F. W. 1952. The effects of rain and temperature on plant distribution in the desert. Proceedings International Symposium on Desert Research, Jerusalem, pp. 232–237.

WENT, F. W. 1957. The Experimental Control of Plant Growth. Chronica Botanica Company, Waltham, Mass.

WENT, F. W. and WESTERGAARD, M. 1969. Ecology of desert plants. III: Developments of plants in the Death Valley National Monument, California. *Ecology* 30:26–38.

WOOD, I. G. 1934. The physiology of xerophytism in Australian plants. *Journal of Ecol.* 22:69–87.

ZAMPIRESCU, N. 1931. Cercetari asupra absorptiunii apei prin organele aeriene ale plantelor. *Bull. Minist. Agr.* 3:3–6.

ZELITCH, I. 1961. Biochemical control of stomatal opening in leaves. *Proc. Nat. Acad. Sci. Wash.* 47:1423–33.

ZELITCH, I. 1963. Stomata and water relations in plants. The Connecticut Agr. Exp. Sta., New Haven, p. 116.

SUBJECT INDEX

Abies alba, 80
Acacia accuminata, 104, 105
Acacia albida,8
Acacia aneurea, 7
Acacia ciliata, 113
Acacia cyanophylla, 36 113, 115, 116
Acacia longifolia, 114
Acacia pendula, 113
Acacia saligna, 113
Acacia spirocarpa, 8, 113, 124
Acacia tortilis, 8, 36, 124
Acclimatization, 16
Acer syriacum, 7
Air borne dust, 58
Air humidity,6
Alfalfa, 54, 97
Alhagi camelorum, 120
Alkali solubility, 15
Aloe, 50
Amides, 109
Amino acids, 109
Amygdalus communis, 12
Anabasis articulata, 8, 113
Anagyris foetida, 12
Anona, 18
Arbutus andrachne, 18
Aridity Index, 36
Artemisia monosperma, 111
Artocarpus integrifolia, 19
Atacom desert, 50
Atmospheric moisture, 49, 129
Atriplex halimus, 8, 13, 14, 112
Atriplex hymenelytra, 108
Avocado, 18
Avvicennia, 19

Balanites aegyptiaca, 114, 124
Bark, 66
Betula, 75, 77, 79, 99
Bicarbonate, 58
Billardiera, scandens, 124
Bimorphism, 111
Biochemical mecanism, 126
Brachyblast, 112
bulk density, 46

Cactaceae, 50, 51

Calcium, 58
Calligonum commosum, 8, 112
Callistemon laurifolia, 114, 124
Callistemon rigidus, 114
Callistemon salignus, 56
Callotropis procera, 8
Cambium, 27, 100, 102
Carbohydrates, 23, 109, 128
Cassia sp., 113
Caatinga shrubs, 112
Cell turgor, 54
Cellulose, 15, 128
Cenomanean, 7
Ceratonia siliqua, 7, 18, 115, 116
Cercis siliquastrum,7
Ceriops, 19
Cheirvanthera linearis, 124
Chloride, 58, 124
Chlorophyl, 128
Citrus paradisii, 115
Citrus sinensis, 54, 56, 115
CO_2, 56, 100, 126
Cocos nucifera, 20
Coffea arabica, 16
Colenchyma, 112
Complex ecosystem, 23
Copparis, 112
Coppicing, 7
Croton floribundus, 112
Crystals, 15
Cucurbita maxima, 18
Cupressus horizontalis, 7, 36, 63
Cupressus pyramidalis, 36
Cuticular transpiration, 127
Cutin, 57, 112

Dead Valley, 50, 59
Deciduous species, 15
dehydration, 94, 120, 127
Dew, 26, 49, 50, 61, 123
Diffuse porous, 15
Disseminating organ, 19
Domesticated, 33
Dominant trees, 106
Douglas fir, 129
D.P.D. 26, 56
Drought years, 36

Dynamic phenomena, 23

Ecotype, 17
Edaphic, 88, 129
Energy of life, 129
Engelman spruce, 89, 129
Enzyme activity, 109, 128
Eocene, 7
Epidermis, 57
Eremophila Sturtil, 125
Erythrosine, 51
Eucalyptus albens, 114
Eucalyptus blakelyi, 114
Eucalyptus camaldulensis, 15, 36, 54
    56, 69, 85, 90, 107, 115, 124, 131
Eucalyptus cinerea, 114
Eucalyptus citriodora, 115
Eucalyptus ficifolia, 114
Eucalyptus gomphocephala, 36, 63, 85
    86, 90, 104, 131
Eucalyptus longifolia, 125
Eucalyptus maculata, 56, 57, 114
Eucalyptus melanophloia, 114
Eucalyptus melliodora, 114
Eucalyptus occidentalis, 36, 54, 56, 85,
    86, 90, 114
Eucalyptus poleanthemos, 105
Eucalyptus pressiana, 105
Eucalyptus pulverulenta, 111
Eucalyptus robusta, 56
Eucalyptus sideroxylon, 114
Eucalyptus staigeriana, 114
Eucalyptus tesselaris, 114
Eucalyptus torquata, 105
Euphorbia sp., 50
European beech, 109
European larch, 109
Eurotya, 14
Evaporation, 65, 76, 83, 87, 126, 130
Evapotranspiration, 45, 46, 47, 76, 130
Evolutionary process, 9
Exotics, 17, 124

Fagus sylvatica, 79, 81, 99, 129
False rings, 102
Fertilizers, 24
Fibrous hull, 20
Firmania, 54, 56
Flowing dew, 62
Foliage, 23, 85
Fuxin, 51

Gaseous exchange, 113
Genista sp., 112
Gleditshia triacanthos, 56

Global radiation, 27, 87
Glycosides, 109
Grapevine seedlings, 119
Grapefruit, 18
Gravimetric methods, 24
Gravitational water, 7, 33
Graya, 14

Haloxylon persicum, 36
Halophyte, 60
Heliotropum rotundifolium, 112
Hemicellulosa, 128
Hexosans, 15
Hoodia lithops, 50
Hydration, 90, 119, 121
Hydrotropism, 119
Hygrophytic, 25
Hypodermal sclerenchyma, 112

Incident solar flux, 131
Index of Aridity, 6, 36
Infrared spectrum, 131
Intermediate sized, 113
Ionic concentration, 59
Irano-Turanian region, 8
Izohyets, 6
Isotopes, 52

Juvenile leaves, 112, 113

Kinetic energy, 9

Larix, 79, 99
Larrea mitida, 120
Latex, 128
Leaf area, 87
Leaf fall, 112
Leaf moisture, 27, 94
Lenticles, 66
Leonardo de Vinci, 50
Light demanders, 128, 129
Lignin, 128
Limestone, 7
Limon, 18
Lithops salicoal, 111
Loma vegetation, 50
Longepole pine, 89

Magnesium, 58
maize, 49, 53
Mangifera indica, 18
Maqui scrub, 130
Medicago sativa, 56
Melaleuca armilaris, 114, 124
Melaleuca styphelioides, 114

Meristem, 21
Mesembryianthemum, 111
Mesophyll, 53
Mesophytic, 25
Metabolism, 16, 23, 117
Metabolic status, 83, 121
Methyl green, 51
Methyl blue, 51
Milky juice, 112
Mist, 44, 61, 123
Moisture content, 33
Moisture depletion, 49
Morphological modifications, 123
Mount Kilimanjaro, 15
Musilagenous materials, 112
Myrtus communis, 8

Namib desert, 50
Needle length, 21
Nerium Oleander, 14
Neutron scatter, 24
Nitrate, 58
Norway spruce, 109

Ochrodenus baccatus, 112, 113
Oils, 24
Olea europea, 11
Oleoresin, 128
Ontogenic variation, 17
Osmotic pressure, 51, 122

Parabola, 84
Parkinsonia aculeata, 113
Pattern of growth, 10
Pattern of stomata, 111
Pelargonium, 50
Penecephylum schottii, 108
Pentosus, 15
Phenylmercuric acetate, 117
Phillyrea media, 7, 115, 124
Photosynthesis, 97, 117, 129
pH, 58
Phragmites communis, 66
Physical evaporation, 57, 75, 117
Physiological modifications, 15
Picea excelsa, 129
Picea engelmanii 80
Pinus brutia, 36, 63, 90, 106
Pinus canariensis, 21, 90
Pinus contorta, 80
Pinus halepensis, 7, 36, 44, 69, 90, 106, 107
Pinus insignis, 102
Pinus pinea, 21, 90
Pinus resinosa, 117
Pinus taeda, 66
Pistacia atlantica, 7, 8, 14, 124

Pistacia lentiscus, 7, 115, 122, 124
Pistacia palaestina, 7, 77, 122, 124
Pittosporum tobira, 56, 115
Pittosporum undulatum, 155
Platanus orientalis, 14
Plasticity, 17
Plastic ducts, 26
Polyethylene, 26, 63
Populus euphratica, 14
Potassium permanganate, 52
Protoplasm, 109
Prosopis species, 113, 120
Prunus amygdalus, 115
Prunus armeniana, 100
Prunus ursina, 7
Pseudotsuga douglassi, 99, 108
Pulpy pericarp, 19
Pyrus malus, 115, 116

Quercus calliprinos, 7, 36, 124
Quercus ilex, 105
Quercus infectoria, 7
Quercus ithaburensis, 7, 12
Quercus pubescens, 97

Radiation energy, 129
Regression techniques, 79
Relative humidity, 65, 75, 84
Rendzina, 48, 102
Resin canal, 53
Rhammus sp., 8, 112
Rhizophora mucronata, 19
Rhus coriaria, 8
Ricinus communis, 11
Rocky Mountains, 80
Root system, 119, 120

Salsola tetranda, 13, 14
Salt crystals, 111
Salvia triloba, 111
Schinus molle, 36, 113, 115
Sclerophilic forests, 124
Sconeratia acida, 19
Scotch pine, 109
Semi arid, 131
Semi desert, 5
Shade bearer, 129
Shamuti, 17
Simondsia californica, 14
Sodium, 58
Soil temperature, 79
Solar energy, 126
Solar illumination, 123
Solar radiation, 125, 129
Sorghum, 109
Spartium junceum, 8

141

Spinoform leaves, 112
Stomatal closure, 56
Stomatal movement, 57
Stomatal opening, 54
Stomatal apertures, 51, 54, 112
Styrax officinalis, 8
Suaeda asphaltica, 112
Suaeda monoica, 113
Succulents, 125
Sulphate, 58
Subterranean leaves, 111
Synecological methods, 23
Synthetic film, 118

Tamarix aphylla, 36, 41, 54, 63, 90, 107, 120, 123
Tamarix pseudopallassii, 36
Tanganyica, 15
Tannin, 15, 24
Temperature, 65, 75, 100, 128
Terra rossa, 7, 36, 47, 107
Theory of phytoplasticity, 16
Termodynamic factors, 44
Thewillea somorae, 51
Thymeleia hirsuta, 36
Thymus capitatus, 111
Topography, 7, 36
Torsion balance, 26
Tracheid formation, 102
Tracheid walls, 100
Transpiration suppressants, 117, 118

Transpirational cooling, 65
Trichocoulen, 50
Trimethylammonium chloride, 117
Trunk, 21
Turanian, 7

Ultraviolet light, 23

Varthemia iphionoides, 111
Vessels, 14
Viscosity-protoplasm, 80
Vitex-agnus-castus, 14

Water deficiency, 57
Water deficit, 57
Water equilibrium, 127
Water molecules, 125
Wheat, 109
White, pine 109
Wilting, 126
Wind velocity, 75, 83, 123, 126
Woody xerophytes, 48

Xeromorphic structure, 113
Xeromorphism, 121
Xerophytes, 49, 120, 122
Xerophytism, 119, 120
Xylem, 15, 128

Zizyphus spina christi, 8
Zygophyllum dumosum, 113